林少宫文集

林少宫 著

华中科技大学出版社
http://www.hustp.com

中国·武汉

作者简介

林少宫　男，祖籍广东信宜，1922年12月出生于北京，2009年11月7日去世，享年87岁。生前任华中科技大学经济学院顾问、特聘教授、博士生导师，数量经济与金融研究中心主任和名誉主任，数量经济学专业学科带头人，曾任华中工学院经济管理学院（现华中科技大学管理学院）院长、数量经济研究所所长。林少宫教授是我国著名的经济学家和统计学家、杰出的教育家，曾担任中国现场统计研究会名誉理事长和中国数量经济学会顾问。在半个多世纪的学术生涯中，他不仅在理论上造诣深厚，出版学术著作10余部，发表学术论文近百篇，为现代经济学特别是计量经济学在中国的发展做出了重要贡献；而且在数理统计的实际应用方面推广了正交实验设计，为国家创造了巨大的经济效益。同时，林少宫教授还"得天下之英才而育之"，培养了数十名优秀的研究生，其中田国强、艾春荣、谭国富、宋敏等，都已经成为知名经济学家，并在国际学术界崭露头角。由于林少宫教授在理论、方法、应用和教学各个方面所取得的杰出成就，其事迹先后被收入《世界（教育界）名人录1987》和《中国世纪专家传略》等传记丛书。获湖北省为改革和建设做出贡献的先进个人（1987）、全国优秀归侨知识分子（1989）和全国优秀质量管理工作者（3次）等称号。华中科技大学专门授予林少宫教授伯乐奖和特聘教授。获国务院政府特殊津贴（1991-2009）。

◆ 1946—1947 年，任上海暨南大学统计学助教。其间，在《经济研究双月刊》发表论文《管子经济思想》，并完成熊彼特（J. A. Schumpeter）等撰写的 Rudimentary Mathematics for Economist（《经济数学梗概》）一书的翻译。

◆ 1948 年 3 月，在美国路易斯安那州立大学（Louisiana State University）完成论文《中国劳工及其运动的简要回顾》。1949 年 5 月，完成硕士学位论文《中国经济制度之历史研究》。

◆ 1949年秋，在美国伊利诺伊大学（University of Illinois）香槟分校开始攻读博士学位。1952年，完成博士学位论文《指数经济理论在真实国民收入中的估值作用》。

◆ 1950年6月，被接纳为美国统计学会正式会员。

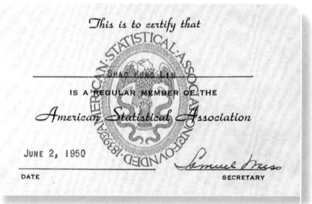

◆ 1952年10月，获美国伊利诺伊大学哲学（经济学）博士学位。

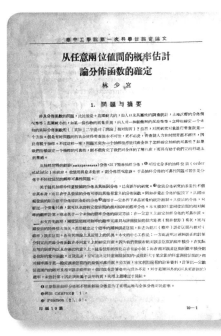

◆ 1957年5月，在华中工学院第一次科学讨论会上发表论文《从任意两位值间的概率估计论分布函数的确定》。

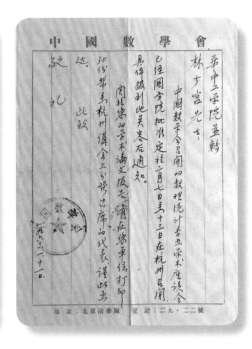

◆ 1963年，参加中国数学会在杭州召开的"数理统计专业学术座谈会"，作题为"序贯分析与应用"的专题报告。同年，著作《基础概率与数理统计》由人民教育出版社出版（被誉为"中国这一学科领域的开山之作"）。

◆ 1973年12月，完成论文《估计缺落数据的交互对比法》，发表于《华中工学院学报（创刊号）》。

◆ 1977—1978 年，林少宫陆续在《湖北农业科学》上发表《正交设计在农业试验上的应用》（1~9 讲，共 14 篇文章），很多读者希望将这些期刊文章印成单行本发行。为此，林少宫做了一些准备工作，他用针线将这些原始文章，加上补充的目录、序言、附录一和附录二，装订在一起，形成单行本原稿，还在临时封面上，写下书名及作者名。

◆ 大约 1979 年，与数学系师生在华中工学院图书馆进行学术交流。

◆ 大约1980年，在野外活动中，与华中工学院院长朱九思及学校同仁合影。

◆ 1980年夏，诺贝尔经济学奖获得者、计量经济模型创建人劳伦斯·R.克莱因（Lawrence R Klein）教授率领美国经济学家代表团与中国社会科学院合作，在北京颐和园举办了为期7周的"计量经济学讲习班"（成为中国数量经济学发展历程中的一个标志性事件）。其间，与克莱因教授夫妇（左、中）合影于颐和园。

◆ 1980年夏，时任中国社会科学院外事局局长王光美同志（右3）来颐和园计量经济学讲习班探班，林少宫（右1）被特邀担任此次讲习班的翻译并讲解，为讲习班的成功举办起到了重要作用。

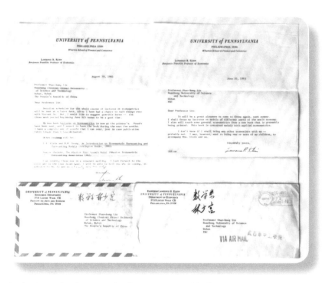

◆ 20世纪80年代，为引进计量经济学，同克莱因教授保持密切联系。1983年6月和8月，克莱因教授在信中说，他将再次来中国，举办为期3周的计量经济学讲座，讲授应用计量经济学、世界经济不同部分的模型等课程。图为与劳伦斯·R.克莱因的来往信件示例。

◆ 1981年2月，与张培刚教授一起完成了《中国的经济发展与外贸前景（详细摘要）》手稿；3月，完成了《中国的经济调整与外贸前景》会议论文；5月，参加了在美国召开的第一届美国与亚洲经济关系会议。

◆ 1981年5月中旬，回母校美国伊利诺伊大学香槟分校访问交流，受到30年前的导师唐纳德·W.帕登（Donald W Paden）教授的热情接待。与帕登教授夫妇在他们自家花园合影留念。帕登教授曾任伊利诺伊大学经济教育中心主任，二战期间任美国商务部的经济学家，在统计学、经济学等领域，颇有建树。

◆ 1981年5月中旬，帕登教授（左）在自家花园与林少宫（中）、张培刚（右）教授合影。

◆ 1982年4月，出任华中工学院数量经济研究所所长。7月，迎来普林斯顿大学周至庄教授的访问，并获朱九思院长的批示："数量经济学十分重要，请多多提出意见。我们下决心大大加强数量经济学的研究，下决心解决有关的具体问题。与此同时，同国外学者的联系十分重要，长期保持。"

◆ 1985年10—11月，访问美国明尼苏达大学，多次与该校著名经济学家、诺贝尔经济学奖得主赫维茨（Leo Hurwicz）教授讨论交流。受赫维茨教授的邀请，做了"中国现代化：稳定、效率和价格机制"专题演讲。

◆ 1985年12月，获清华大学经济管理学院首任院长的感谢和祝福。

◆ 1986年1月，出任华中工学院经济管理学院（现华中科技大学管理学院）首任院长。任内，对管理工程博士学位授予权的申报起到重要作用。

◆ 1986年11月上旬，在日本参加第二届中日统计讨论会，发表论文"Should Finite Population Inference Follow a Different Principle?"。前排右4为作者。

张龙翔教授和林少宫教授　1986年11月

◆ 1986年11月中旬，与北京大学前校长张龙翔教授等应邀前往日本创价大学访问。访问期间，和张龙翔教授分别为该校师生做了技术进步测度和中国教育改革的演讲。

◆ 1987年6月29日上午，在武汉华中工学院主持了"中美经济合作学术会议"开幕式。这是我国首次在高校召开的国际经济讨论会。前排右5为帕登（D Paden）教授，右6为张培刚教授，右7为杜塔（M Dutta）教授，右11为朱九思院长，左1为作者。

◆ 1987年6月30日媒体关于"中美经济合作学术会议"的报道。

◆ 1987年7月，作者当年的导师帕登（D Paden）教授（中）在中美经济合作学术会议上宣讲了论文"Technology and Formation of Human Capital"，并应邀共同主持会议的专题讨论。右为作者。

◆ 1987年7月，中美经济合作学术会议结束后，为杜塔（M Dutta）教授饯行。左2、4、5、6分别为张培刚、杜塔、吴驯叔、林少宫教授，右1为杜塔夫人。

◆ 1988年4月，与夫人吴驯叔教授合影于华中工学院校园。吴教授为首届"中美经济合作学术会议"的筹备做出了重要贡献。

◆ 1991年5月17日，与夫人吴驯叔教授在学校图书馆前合影。

◆ 1992年6月25日，与数量经济学专业1988级全体学生合影，2排左5为作者，左6为吴燮和教授，左7为李楚霖教授。

◆ 1993年，朱九思老校长会见了回母校讲学的田国强教授（右2），同年，由田国强教授主编的《市场经济学普及丛书》（14本）问世，其当年导师林少宫教授（右1）和李楚霖教授（左1）为此撰写了《简明经济统计与计量经济》一书。

◆ 1994年3月14日，在《湖北日报》"为荆楚崛起献良策"专刊上，报道了"华中理工大学教授、经济学家林少宫（左），为武钢冷轧产品提高质量做出过重大贡献。目前，他和李楚霖（右）教授正从事证券市场完善性检验的研究"。/《湖北日报》记者郑元昌摄影报道

◆ 1998年1月31日，与来访的弟子宋敏教授合影。

◆ 2000年6月，与诺贝尔经济学奖得主赫维茨（Leo Hurwicz）教授（右2）、校长周济院士（右1）和田国强教授（左1）亲切交谈。

◆ 2000年6月13日，出任华中科技大学数量经济与金融研究中心主任。

◆ 2001年6月,在"林少宫教授八旬华诞暨从教50周年庆贺会"上发言。

◆ 2001年12月15日,湖北省现场统计研究会第四届年会上,举办了庆祝名誉理事长林少宫教授八十寿辰暨从教五十年活动,代表们一致向林老致以热烈的祝贺,并为他对促进数理统计、正交试验、数量经济等学科在我国的发展做出突出贡献而赞誉。作者在会上作了重点发言。

◆ 2002年6月5日,与诺贝尔经济学奖得主麦克法登(Daniel McFadden)教授握手,在场的有王乘副校长(中)、艾春荣教授(左)和徐长生院长(右)等。

◆ 2002年6月5日，与诺贝尔经济学奖得主麦克法登、学院同事邓世兰（左1）、唐齐鸣（左2）、徐长生（右2）和张建华（右1）合影。当天，麦克法登教授和林少宫教授等共同发起设立"麦克法登－林少宫经济学奖学金"。

◆ 2002年9月18日，与来访的电子工程专家黄秉聪合影。来访者在照片背面写道："尊敬的林教授，您'吃的是草，挤出的是奶'！共和国幸有您一样的精华！学生为有您一样的师长而骄傲。秉聪02.12于加州"。

◆ 2003年6月，人民网、湖北日报、中国教育报和科技日报等10多家媒体报道了"麦克法登－林少宫经济学奖学金"首次颁奖的消息。

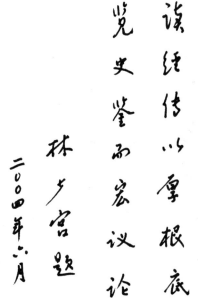

◆ 2004年6月，为华中科技大学经济学院10周年院庆题词。

◆ 2008年12月，作者上了《华中大导师》期刊的封面。"经世良师 林少宫先生生活简朴，为人低调，不求闻达，耕耘喻园五十余年，甘受其默。然所谓桃李不言，下自成蹊，凡知先生品行学问者，谁不交口称赞？"

◆ 2009年11月16日《华中科技大学周报》（2009年11月16日第01版）报道："学习钱学森贝时璋林少宫先生精神座谈会召开"，校党委书记路钢出席座谈会并殷切寄语与会教师代表。

◆ 2010年11月7日，林少宫教授逝世一周年，经济学院院长徐长生教授（左3）、党总支书记邓华和教授（右2）、副院长唐齐鸣教授（左2）、刘海云教授（右1）以及林少宫教授学生代表徐慧玲（左1）等经院师生赴九峰山为林少宫先生扫墓。

◆ 2019年12月16日上午，林少宫纪念讲座揭幕式暨第一次讲座在华中科技大学经济学院报告厅隆重举行。出席活动的嘉宾有林少宫的夫人吴驯叔教授、长子林子美先生，上海财经大学高等研究院院长田国强教授，武汉大学经济与管理学院院长宋敏教授，华中科技大学经济学院党委书记江洪洋、院长张建华及学院师生代表。前排左起：田国强，吴驯叔，宋敏。后排左起：曹小勇，王红，张建华，林子美，徐长生，唐齐鸣，江洪洋，欧阳红兵。

◆ 2021年11月7日，华中科技大学经济学院隆重举行林少宫纪念讲座，主讲嘉宾为香港中文大学（深圳）校长讲座教授艾春荣、助理教授陈齐辉。张建华院长表示此讲座旨在纪念杰出的数理统计学家、计量经济学家林少宫先生对我国计量经济学和我校的卓越贡献，旨在铭记老一辈学者的报国情怀和对学术孜孜不倦的追求精神；激励一代代学子积极传承家国情怀，领悟经世济民的思想，将个人的发展与国家、民族的命运联系在一起，为国家的现代化建设贡献自己的力量。原院长徐长生、书记邓世兰和邓华和，唐齐鸣、杨继生教授等学院师生代表参加了此次讲座。左起：杨继生，王红，邓世兰，张建华，艾春荣，唐齐鸣，陈齐辉，欧阳红兵，沈淑琳。

内容简介
Abstract

本书主要梳理了林少宫教授学术生涯中在数理统计、正交试验设计、数量经济、国际交流方面的学术成果。林少宫教授在数理统计学领域的研究填补了该学科的空白，并成为之后相关研究的一个模板；在正交试验设计领域的研究与应用取得丰富成果，属国内首创，独具中国特色。现遴选林少宫教授的部分著作和论文出版成集，旨在梳理林少宫教授学术生涯的成果，为从事计量经济学研究的学者、学生指明方向。

林少宫教授是我最敬爱和最敬重的恩师！老师和我虽说是师生关系,但情若父子、亦师亦友,可以毫不夸张地说,没有老师在德和学方面的培育和潜移默化的身教,就没有今天作为经济学家的我。多年前,我曾先后撰写《感念师恩——记我的导师林少宫教授》(2007)和《悼吾师林少宫教授》(2009)两篇文章追忆老师。今年适逢恩师逝世十周年,母校华中科技大学(以下简称华科)经济学院有心启动了《林少宫文集》的编辑,收录了包括老师的博士论文在内的许多珍贵文献,其中一些未在国内出版过,令我十分欣慰和感动！华科经济学院院长张建华教授和恩师的长子林子美先生嘱我为文选作序,我义不容辞地接受了这一任务。

　　与老师相识是在1979年,那年我报考了他的研究生而结下师生情缘。自那时起,我就一直与老师保持了长达30年的非常密切的联系,直至他2009年87岁高寿离世。我1983年出国留学、从教之后,我们之间还经常通信,直到更为方便的打电话和发电邮时才停止写信,至今我保存着老师写给我的几十封信件。在1988年2月18日给我的一封信中,老师情真意切地说:"在国内要创办一些专业,还是困难多端,一言难尽,但无论如何都应为祖国振兴而效力。"如今"无论如何都应为祖国振兴而效力"这句话,已经被镌刻在华科经济学院大楼内老师的雕像上,也被放在这本文集的扉页上。我想,这句话既是老师对我的期许,也是他本人的自勉,贯穿了他的学术生涯。

　　老师1922年出生,1944年毕业于中央大学经济系,1947年赴美留学,但他保持着对中国发展的密切关注,次年就用英文撰写并发表了"A Brief Review of Chinese Labor and Its Movement",对中国的劳工及其运动向国外做了介绍。老师先后在路易斯安那州立大学和伊利诺伊大学接受了系统的现代经济学训练,分别获得硕士和博士学位。也正是在伊利诺伊大学攻读博士学位期间,他与在该校任职的赫维茨(Leo Hurwicz)教授也是我后来的博士生导师相识相知。赫维茨在2008年给老师的一封信中写到,他对林老师对经济理论认知的敏锐性印象深刻。在那里,老师也结识了另一位诺贝尔经济学奖得主莫迪利亚尼(Franco Modigliani),他将留学时期的老师描述为"从中国到来不久的一位精明而求职心切的学生""一名很好的学生",认为与老师"有着许多共同的有趣经验"。

1952年,老师从伊利诺伊大学博士毕业,博士论文题为"The Economic Theory of Index Numbers in the Valuation of Real National Income"。我首次接触到老师的博士论文时,让我惊讶的是以数理统计研究著称的他,博士论文竟然是关于经济理论的,其中运用了大量的数学工具和当时的最前沿经济理论。从个人偏好出发来分析总和(aggregation)的社会福利变化问题,这在当时是非常前沿的研究,直至现在也依然是经济学中的难题。老师博士论文中这种融数学分析于经济分析的方法在"二战"之后的现代经济学中才真正开始兴起和迅速发展。同时,他紧跟当时的学术前沿,在老师论文的参考文献中,引用了大量 American Economic Review、Econometrica、Review of Economic Studies 等顶尖期刊的最新论文,其中不乏希克斯、萨缪尔森、阿罗、克莱因、库兹涅茨、里昂惕夫等后来的诺贝尔经济学奖得主,也包括凯恩斯、庇古、费雪、霍特林等经济学名家的大作。难怪老师得到了经济学大师霍特林的欣赏,在美国俄亥俄州的戴顿大学任教期间就被霍特林力邀到所任教的北卡罗来纳大学教堂山分校(University of North Carolina at Chapel Hill)去工作,但老师毅然选择了回国。我不禁感叹,如果当时老师继续留在美国从事经济学研究的话,也许他的学术生涯会另有一番新的天地。

1953年,老师受聘到美国俄亥俄州的戴顿大学任教,讲授统计学、经济学原理和美国经济史。也许是巧合,这些课程与熊彼特所言的科学的经济分析三要素——统计、理论和经济史几乎完全一致。的确,理论逻辑、实践真知和历史视野的结合是确保学术研究的科学性、严谨性、现实性、针对性、前瞻性、思想性的关键。阅读本文集中老师的论文,包括1954年回国之后所发表的研究成果,我们也可以看到上述"三维六性"的良好学术呈现,以及一种中西汇通、比较的治学方法。也正是这种训练所打下的坚实基础,使得老师在20世纪70年代末能够很快地重拾学术研究,并迅速与国际学术研究前沿接轨,和国外学者交流,让我们这些学生无论是学习还是出国深造都受益。老师的开放国际学术视野为我们这些学生在改革开放之初的经济学学习打开了一个广阔的空间。

我、徐滇庆、方振民、孙兆伦、艾春荣、宋敏、谭国富等是老师在20世纪80年初最早的两届学生,他就要求我们研读经济学领域非常前沿的文献和当时最流行的教科书和前沿专著,如 Akira Takayama 的著作 *Mathematical Economics*, Franklin Fisher 的专著 *Identification Problem in Econometrics*, John S. Chipman 的关于总和问题的文章。而在我出国之前,我的硕士论文就是学习了老师从国外访学带回的前面所提及的 Franklin Fisher 教授的专著后而撰写的,完成之后我将硕士论文寄给了该书作者也是 *Econometrica* 主编的 Franklin Fisher 教授,获得了他的好评,并收到了长达三页的评论,这使我对做研究的信心大增,对我后来赴美留学深造也起到了极大的促进作用。一个以理工见长的高校,华科能够孕育培养出一大批在国际学术舞台享有盛誉的经济学家,得益于林少宫、张培刚等具有世界眼光的经济学家、教育家,也得益于当代真正的教育家和改革者朱九思老校长。

老师对朱九思老校长非常推崇,而九思老校长对老师的贡献也念兹在兹。在2004年的一次大会上,九思老校长曾提及,20世纪70年代末华科做了两件全国第一的改革工作:一件是将原来单一的工科院校向综合性大学转变;另一件是在全国率先建立了数量经济学专业。老师正是这个专业的旗帜!他一生出版10余部计量经济学方面的经典著作、译著,为计量经济学在中国的发展做了大量奠基性贡献。他曾提及,要解决数量经济学研究中存在的问题,"根本办法还在于教育。我是寄希望于年轻一代的。我们要教会年轻一代切勿脱离中国经济的实际,用你在微、宏观经济学中学到的理论分析框架,建立你所研究问题的模型,然后寻找最先进的计量方法去计算。"可谓一语中的!

2004年以来,我在上海财经大学进行的全方位经济学教育改革,在很大程度上也正是得益于老师和朱九思老校长给我的引路、启发。尤其是老师之于经济学的理论创新及在中国的学科建设和人才培养,具有筚路蓝缕之功。其实不仅在上述这些方面,在理论的实际应用方面老师也做出了积极贡献,仅他所倡导的正交试验设计一项的研究和推广就产生了巨大经济效益。自20世纪70年代起,老师研究和推广应用正交设计成果(本选集也收录了其中2篇论文),编制了"正交试验极差临界值系数表",大大简化了正交试验的分析和计算工作,加速了在我国工农业生产试验中的广泛应用,在1994年统计研究会的新闻发布会上被评估为创造了30亿元以上的经济效益。

古语讲,人生三不朽,立德、立功、立言!老师在立德做人、立功做事、立言做学问这三个方面都堪称表率。本文集所收录的文章言得其要、理足可传,它的出版可以让我们有机会一睹老师的学问之道与学术脉络,领略老师的理论贡献与学术创见。

是为序!

上海财经大学高等研究院院长、教授
2019 年 9 月 24 日
修改于 2021 年 11 月 8 日

目录
Contents

第一部分　早期研究

The Economic Theory of Index Numbers in the Valuation of Real National Income　/3
  1. Introduction　/3
  2. The Valuation of an Individual Consumer's Real Income　/4
  3. The Valuation of Real Income for a Group of Individuals　/18
  4. The Valuation of Real Income for a Group of Individuals (continued)　/45
  5. Imperfection of the Economy and Valuation Problems　/52
  6. Summary and Conclusion　/68
  7. Appendix A: A Note on Fisher's Concept of Real Income　/72
  8. Appendix B　/73
Acknowledgement　/75
VITA　/75

第二部分　数理统计

估计缺落数据的交互对比法　/79
  1. 前言　/79
  2. 多个因素水平的复杂性　/80
  3. 对比、交互对比　/82

4. 交互对比的一般形式　　　　　　　　　　　　　　　/85
　　5. 最小二乘法　　　　　　　　　　　　　　　　　　　/89
　　6. 在拉丁方上的应用——内加条件法　　　　　　　　　/94
　　7. 在部分析因实验上的应用　　　　　　　　　　　　　/98
　　8. 结语　　　　　　　　　　　　　　　　　　　　　　/101
近代数理统计的精巧性和应用性　　　　　　　　　　　　　/103
　　1. 什么是统计推断？　　　　　　　　　　　　　　　　/103
　　2. 矩法并不总是合理的　　　　　　　　　　　　　　　/104
　　3. 凭什么进行估计？　　　　　　　　　　　　　　　　/104
　　4. 显著性判断，不是乱猜？　　　　　　　　　　　　　/107
　　5. Fisher，Wald，各有所执　　　　　　　　　　　　　/108
抽样调查的代表性问题及有关理论和概念的历史发展　　　　/110
　　1. 引言　　　　　　　　　　　　　　　　　　　　　　/110
　　2. 调查样本设计　　　　　　　　　　　　　　　　　　/110
　　3. 估计理论　　　　　　　　　　　　　　　　　　　　/115
　　4. 结束语：理论与实践　　　　　　　　　　　　　　　/117
怎样在加速寿命试验中构造置信区间　　　　　　　　　　　/119
　　附件：给林少宫教授的联名信(1988年3月31日)　　　/123

第三部分　正交试验设计

正交试验设计讲座(农机试验例解)　　　　　　　　　　　/127
　　1. 什么是正交试验设计　　　　　　　　　　　　　　　/127
　　2. 怎样做正交试验　　　　　　　　　　　　　　　　　/128
　　3. 正交设计的原理　　　　　　　　　　　　　　　　　/131
　　4. 多指标问题　　　　　　　　　　　　　　　　　　　/132
　　5. 交互作用的考察　　　　　　　　　　　　　　　　　/133
　　6. 正交表的使用　　　　　　　　　　　　　　　　　　/136
正交设计法　　　　　　　　　　　　　　　　　　　　　　/138
　　1. 概述　　　　　　　　　　　　　　　　　　　　　　/138
　　2. 区组试验(局部控制)　　　　　　　　　　　　　　　/139

   3. 正交拉丁方与正交表 /141
   4. 析因试验及其部分实施 /145
   5. 正交表中交互作用列的表示与构造 /153
   6. 部分析因中的效应混杂 /158
   7. 区组混杂 /162
   8. 多因素优选法 /166
   9. 几种常用正交表 /175
   10. "正交设计法"相关和反馈信息 /181

第四部分　数量经济

正交实验设计在微观计量经济中的应用——政策（或事件）评价法
                                                              /191
   1. 回归设计与分析 /191
   2. 正交设计与分析 /193
   3. 正交设计的价值 /194
   湖北省现场统计研究会第四次代表大会概况 /195
多元线性回归系数的"其他情况不变"释义 /196
   1. 引言 /196
   2. 多元回归的 OLS 估计与解释 /197
   3. 受控变量过少的严重性 /198
   4. "代理"与"工具" /200
   5. 讨论 /201
   6. 因果效应与综列数据 /204
   7. 结束语 /205
微观计量经济学中的实验设计法——期望"计量经济（学）设计与
分析"的出现 /206
   1. 控制什么？怎样控制？ /207
   2. 不可观测变量的控制 /208
   3. 非严格的可估计函数 /209
   4. 实验经济学 /210
   5. 期待"计量经济（学）设计与分析"的出现 /211

第五部分　国际交流

Aspects of Technology Transfer:China's Experiences　　/215
  1. Economic Growth: The End Product of Technology　　/215
  2. Technology Transfer:Wage, Employment and income Distribution Profile　　/219
  3. Short-Run Versus Long-Run:The Familiar Economic Debate　　/221
  4. Technology Transfer and Product Quality　　/223
  5. Free Trade Regime and Technology Transfer　　/225
  6. Cooperation and Competition　　/228
  7. Conclusion　　/229
  Note　　/229

Should Finite Population Inference Follow a Different Principle?　　/230
  1. The Challange　　/230
  2. Doubts about the MVU Principle and "Therefore"　　/231
  3. Relevance of the Identifier　　/232
  4. Fundamentals of Neyman's Principle Revisited　　/233
  5. A Practical Viewpoint　　/234

附件：在日本创价大学的演讲(1986 年 11 月 11 日)　　/235

China's Modernization:Stability, Efficiency, and the Price Mechanism　　/238
  1. Stability,a Priority　　/238
  2. A Retreat From Efficiency　　/240
  3. Formation of Price in a Socialist Economy　　/243
  4. Reform of the Irrational Price System　　/246
  5. Avoiding Inflation and Ensuring Success　　/249
  6. Appendix:Chronology in the Price History of China　　/253

Acknowledgment /255
Notes /255

参考文献 …………………………………………… 258

后记 ………………………………………………… 269

第一部分
早期研究

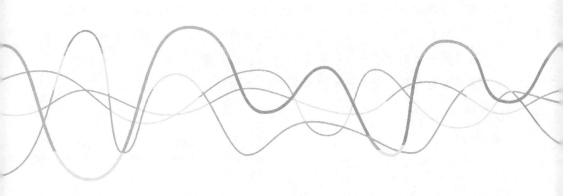

20世纪40—50年代,林少宫教授撰写了多篇重要经济学论文,研究广泛,从中国古代经济思想和制度到当代经济学,包括 *GUANTZE Economic Thought*(管子经济思想,1947年)、*A Brief Review of Chinese Labor and Its Movement*(中国劳工及其运动的简要回顾,1948年)、*A Historical Study of Chinese Economic System*(中国经济制度之历史研究,硕士论文,1949年)和 *The Economic Theory of Index Numbers in the Valuation of Real National Income*(指数经济理论在真实国民收入中的估值作用,博士论文,1952年)等。其博士论文被《美国经济评论》分类为计量经济与统计学,其论文导师唐纳德·W.帕登(Donald W Paden)教授给出高度评价:"相对其后来所处的环境而言,他堪称伊利诺(ILL.)大学的一位杰出学生。"因这篇论文,1954年林少宫先生获当代美国统计学界、经济学界大师哈罗德·霍特林(Harold Hotelling)教授的工作邀请,但他谢绝了邀请,于当年毅然回到中国。攻读博士学位期间,林教授有幸结识了尔后的诺贝尔经济学奖得主赫维茨(L. Hurwicz)教授和莫迪利亚尼(F. Modigliani)教授,多次亲聆其授业,也经常从考尔斯(Cowles)委员会的学术活动中受益,以至他后来把博士论文写成计量经济与统计学一类的论文。仅从参考文献,可看出他跟踪当时的学术前沿,引用了大量顶尖期刊的最新发表的论文,其中不乏希克斯、萨缪尔森、阿罗、克莱因等后来的诺贝尔经济学奖得主,也包括凯恩斯、庇古、费歇尔、霍特林等经济学名家的大作。这篇论文被本章收录,其他的因篇幅所限未列入。

The Economic Theory of Index Numbers in the Valuation of Real National Income[①]

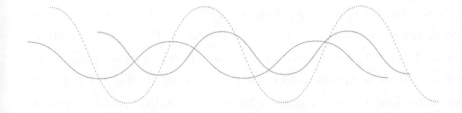

1. Introduction

The valuation of real national income is a comparison of the nation's economic welfare between two or more situations rather than a measurement of it in any one situation. Economic welfare in any one situation is dependent on the goods and services produced in that situation. Comparison with other situations calls for the construction of quantity index numbers. It is the purpose of the present study to examine fully the meaning and use of aggregate quantity index numbers in making comparisons of this nature.

Due to recent developments in the theory of consumer's choice, indifference analysis can be used to demonstrate the relations between "true" index numbers and Laspeyres' and Paasche's index members. From these relationships the individual consumer's real income may be evaluated effectively. Indifference analysis of the aggregate quantities for a group of individuals, however, is complicated by considerations of income distribution.

A change in the distribution of income usually leaves some individuals in a better position and others in a worse position. Under these

---

① 1952 年在美国伊利诺大学香槟分校完成的博士论文。

circumstances, any interpretation of changes in <u>actual</u> real national income involves a dubious interpersonal comparison. On the other hand, in making comparisons of <u>potential</u> real national income one must consider the changes in collective demand and prices, as well as the change in production resulting from a redistribution of income.

Government services and producers' goods——obviously constituting a part of the total product——must be considered along with consumers' goods in the valuation of real national income. However, while consumers' goods are valued according to their marginal utilities, producers' goods are valued according to their marginal productivities and government services are valued at cost. With goods and services of different sectors of the economy valued differently, no simple, consistent, interpretation of the resultant aggregate quantity index is possible. Thus, unless uniformity of valuations is attained, further restrictions on the use of aggregate quantity index numbers as interpretative devices must be kept in mind.

Despite the fact that voluminous literature can be found on the subject of index numbers, contributions to the economic theory of index numbers have been fragmentary and scattered. Moreover, discussions are largely limited to an individual consumer or its equivalent (such as the representative member of a group or a group of individuals with like tastes and income). Theoretical inquiries into the use of index numbers in the valuation of real national income have been sporadic and incomplete. In this study, a thorough investigation is undertaken of the economic theory of index numbers insofar as this is related to the valuation of real national income. Needless to say, the inherent difficulties in the valuation of real national income necessarily lead to divergency of opinion. The present study is limited to demonstrating exactly what can be known about changes in real national income through the use of index number criteria alone.

2. The Valuation of an Individual Consumer's Real Income

A nation is made up of individuals. Real national income is not a separate entity but consists of individuals' real income. A valuation of real national income is therefore a valuation of the real income received by all the individuals who form the nation. Before we can make a valuation of the nation's real income, we must know first how an individual's real income can be valuated. An individual's real income is measured by the amount of utility yielded by the goods (and services) which he receives. As a preliminary step, let us consider an individual who receives no other goods than those purchased from the market for immediate consumption. As the individual under consideration purchases different bundles of goods in different situations, how can we compare them in terms of utility?

2.1 Basic Assumptions

In order to make possible such comparison of utility, a number of assumptions are required.

(1) The individual's tastes are constant. The same bundle of goods will yield to him the same amount of satisfaction in all the situations to be compared. Technically speaking, he has a fixed indifference map.

(2) The same variety of goods are available in either of the situations to be compared. Every good available in one situation has its equivalent in other situations.

(3) The allocation of his expenditures is an optimum such that the marginal utilities of the goods which he purchases are proportional to their prices. Or put otherwise, the marginal rate of substitution between any two goods is equal to their price ratio. Thus, for the amount of money he spends the amount of utility he obtains is a maximum; for the amount of utility he obtains the amount of money he spends is a minimum.

(4) Utility is an increasing function of quantities of goods. More goods are better than less. Prices being the same, the greater the amount of money spent, the greater the amount of utility obtained.

2.2 Notations

The following notations, alphabetically arranged, will be used throughout this study unless otherwise indicated:

$a_1 \equiv (a_1^1, a_1^2, \cdots, a_1^r)$; $a_2 \equiv (a_2^1, a_2^2, \cdots, a_2^r)$; $\cdots$ the r productive factor prices in Situation Ⅰ, Ⅱ, $\cdots$.

$D_1, D_2, \cdots$ the distribution of purchasing power in Situation Ⅰ,

II , ⋯.

E (q_1)   the level of productive efficiency corresponding to the production of q_1.

E (q_2)   the level of productive efficiency corresponding to the production of q_2.

$L_p = \sum p_2 q_1 / \sum p_1 q_1$ the Laspeyres' price index.

$L_p = \sum p_1 q_2 / \sum p_1 q_1$ the Laspeyres' quantity index.

$\overline{L}_p = \sum \overline{p}_1 q_2 / \sum \overline{p}_1 q_1$ the Laspeyres' quantity index with the virtual (equilibrium) prices as weights.

$p_1 \equiv (p_1^1, \cdots, p_1^s); p_2 \equiv (p_2^1, \cdots, p_2^s); \cdots$ the prices of the s goods in Situation I , II , ⋯.

$\overline{p}_1 \equiv (\overline{p}_1^1, \overline{p}_1^2, \cdots, \overline{p}_1^s); \overline{p}_2 \equiv (\overline{p}_2^1, \overline{p}_2^2, \cdots, \overline{p}_2^s); \cdots$ the virtual prices of the s goods in Situation I , II , ⋯.

$p_1^1 = \overline{p}_1 - p_1; p_2^1 = \overline{p}_2 - p_2; \cdots$.

$p_P = \sum p_2 q_2 / \sum p_1 q_2$ the Paasche's price index.

$p_P = \sum p_2 q_2 / \sum p_2 q_1$ the Paasche's quantity index.

$\overline{p}_P = \sum \overline{p}_2 q_2 / \sum \overline{p}_2 q_1$ the Paasche's quantity index with the virtual prices as weights.

$q_1 \equiv (q_1^1, q_1^2, \cdots, q_1^s); q_2 \equiv (q_2^1, q_2^2, \cdots, q_2^s); \cdots$ the quantities of the s goods in situation I , II , ⋯.

q^* the individual's quantities to be distinguished from the aggregate quantities whenever necessary.

$\hat{q}^*$ the redistributed individual's quantities.

$\overline{q}_1 = q(U_1; p_2)$ the bundle of goods which would have been bought at p_2 prices such that $U(\overline{q}_1) = U(q_1)$.

$\overline{q}_2 = q(U_2; p_1)$ the bundle of goods which would have been bought at p_1 prices such that $U(\overline{q}_2) = U(q_2)$.

$T_p(U_1) = \sum p_2 \overline{q}_1 / \sum p_1 q_1$ the "true" price index at the level of real income $U(q_1)$.

$T_p(U_2) = \sum p_2 q_2 / \sum p_1 \overline{q}_2$ the "true" price index at the level of real income $U(q_2)$.

$T_q = U(q_2) / U(q_1)$ the "true" quantity index.

$U(q_1)=U_1$ the amount of utility yielded by q_1.
$U(q_2)=U_2$ the amount of utility yielded by q_2.
$U(\bar{q}_1)$ the amount of utility yielded by $\bar{q}_1$.
$U(\bar{q}_2)$ the amount of utility yielded by $\bar{q}_2$.
$v=\sum p_2 q_2 / \sum p_1 q_1$ the value (of expenditure) index.
$x_1 \equiv (x_1^1, x_1^2, \cdots, x_1^r); x_2 \equiv (x_2^1, x_2^2, \cdots, x_2^r); \cdots$ the r productive factor quantities in Situation I, II, ⋯.

2.3 The "True" Index Numbers

"True" price index numbers have received constant attention for the past thirty years.[1] As contrasted with index numbers which indicate the amounts of expenditure in different situations in purchasing a given bundle of goods, "true" index numbers of price indicate the minimum amounts of expenditure in different situations for obtaining the same amount of utility or real income. By analogy, a quantity index number may be defined as "true" which is not supposed to show the relative amounts of expenditure in purchasing different bundles of goods at given prices, but the relative amounts of utility yielded by different bundles of goods. In other words, the "true" quantity index number is a utility index number.

Generally speaking, every commodity is more or less substitutable. From the law of diminishing marginal utility we derive the law of diminishing marginal rate of substitution. A consumer will maximize his utility by equalizing the marginal rate of substitution between any two commodities and their price ratio. As relative prices change, substitution takes place. For this reason, neither the Laspeyres' formula nor the Paasche's will give us the "true" index numbers because, as will be seen, both err in assuming that the same amounts of goods will be brought regardless of the change in relative prices.

In any practical situation "true" index numbers can be obtained only with extreme difficulty. Our effort will be mainly directed to the exploration of the relationships among the Laspeyres' and the Paasche's

---

[1] For historical interest, see Frisch, Ragnar. Annual Survey of General Economic Theory: The Problem of Index Numbers[J]. Econometrica, 1936, 4(1): 10-13.

index numbers and the "true" index numbers.

2.4 Indifference Analysis of the Limits

For the sake of simplification, let there be only two commodities X and Y. In Figure 1[1], let the individual consumer possess h units of X and k units of Y in the initial position.

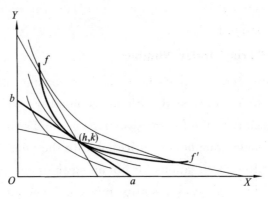

Figure 1

Where the price ratio between X and Y is oa/ob. He is in equilibrium because the point (h,k) is a point of tangency between ab and his indifference curve (the lighter one). It is clear that this point is superior to any other point on or below ab; the latter may be called the price or the expenditure line. As the price ratio changes, he can always move to a higher indifference curve by making substitution between X and Y with his initial possession, or what is the same thing, with the amount of expenditure just enough to buy h units of X and k units of Y at the new price ratio. The locion which he moves as the price ratio changes form a continuous "offer curve" $f\ f'$ every point on which is tangential to a different pair of indifference curve and expenditure line.[2] Thus, every point on or above the offer curve represents a better position than the initial one.

---

[1] This figure is taken directly from Samuelson P A. Foundations of Economic Analysis[M]. Cambridge: Harvard University, 1948:148.

[2] For further description of "offer curve," see Schultz Henry. The Theory and Measurement of Demand[M]. Chicago: University of Chicago, 1937: 25; Bowley A L. The Mathematical Groundwork of Economics[M]. Oxford: Clarendon, 1924:5-8.

The above relations of the points in Fig. 1 are the ones that can be seen from index numbers. Let the point (h, k) be designated as q_1, representing the bundle of goods in Situation I. Thus, oa/ob corresponds to p_1 relative prices. For purposes of comparison we now select another point to represent q_2 in Situation II. If a point on or below ab is selected as q_2 (Fig. 2a), the line $a'b'$ drawn through q_2 parallel to ab must either coincide with or lie below ab. Hence, $oa \geq oa'$ and $ob \geq ob'$. Substituting either the expenditures which oa and oa' represent (in terms of X) or the expenditures which ob and ob' represent (in terms of Y), we have $\sum p_1 q_1 \geq \sum p_1 q_2$. Since every point on $a'b'$ (representing the same amount of expenditure at p_1 prices) must be either on or below ab, we also know that $U(q_1) > U(q_2)$ if $\sum p_1 q_1 \geq \sum p_1 q_2$.

On the other hand, if any point on or above the offer curve ff' is selected as q_2 (Fig. 2b), and through q_2 we draw the new expenditure line cd showing what combination of goods could be bought for the same amount of expenditure at p_2 prices (oc/od), then any point on or below cd (except q_2) is inferior to q_2. Through q_1 we draw $c'd'$ parallel to cd. Since q_2 is a point on or above ff', $c'd'$ must either coincide with or lie below cd. Hence $oc \geq oc'$ and $od \geq od'$. Substituting the expenditures represented by oc and oc' (or od and od'), we have $\sum p_2 q_2 \geq \sum p_2 q_1$. Since every point on $c'd'$ (representing the same amount of expenditure at p_2 prices) must be either on or below cd, we also know that $U(q_2) > U(q_1)$ if $\sum p_2 q_2 \geq \sum p_2 q_1$.

Though we may draw through q_2 in Figure 2b $a'b'$ parallel to ab and find that $a'b'$ lies outside ab, the mere fact that $a'b'$ lies outside ab or, in terms of index numbers, $\sum p_1 q_2 > \sum p_1 q_1$ does not ensure that every point on ab (representing the same amount of expenditure at p_1 prices) is inside cd, i.e. $U(q_2) > U(q_1)$. Similarly, in Figure 2a, we have no assurance that every point on cd is inside ab: $\sum p_2 q_1 > \sum p_2 q_2$ does not ensure $U(q_1) > U(q_2)$.

Thus, $\sum p_2 q_2 \geq \sum p_2 q_1$ serves as a limit in the sense that at least income has increased from Situation I to Situation II, and $\sum p_1 q_1 \geq \sum p_1 q_2$ also serves as a limit in the sense that at least real income has decreased

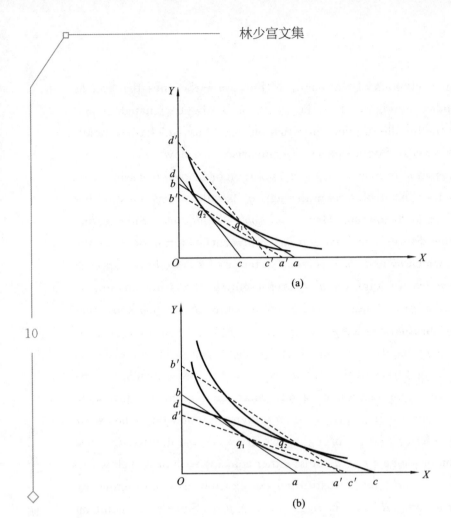

Figure 2

from Situation Ⅰ to Situation Ⅱ. It is possible that $U(q_2) > U(q_1)$ when $\sum p_2 q_2 < \sum p_2 q_1$ and $U(q_1) > U(q_2)$ when $\sum p_1 q_1 < \sum p_1 q_2$, but we cannot be sure. If our knowledge is restricted to the two bundles of goods actually purchased at the two expenditure lines ab and cd, we are ignorant of any point falling within $aq_1 c'$ as compared with q_1 in Figure 2b and any point falling $bq_1 d'$ as compared with q_1 in Figure 2a. This region of ignorance is larger the greater the change in the price ratio and hence the greater the possibility that $U(q_2) > U(q_1)$ when $\sum p_2 q_2 < \sum p_2 q_1$ and $U(q_1) > U(q_2)$ when $\sum p_1 q_1 < \sum p_1 q_2$. We see also from Figure 1 that if the condition $\sum p_2 q_2 = \sum p_2 q_1$ is met, the greater the change in relative prices, the greater the amount by wnich $U(q_2)$ exceeds $U(q_1)$.

It is hardly necessary to add that though in both Pq and Lq the quantities are weighted by the prices which represent the marginal utilities of the goods in Situations Ⅱ and Ⅰ respectively, what we arrive at is a comparison of total utility. But, just because we know nothing from the price and quantity data but the relative magnitudes of the marginal utilities of these goods, we are not able to compare $U(q_2)$ and $U(q_1)$ on all occasions. In order to compare $U(q_2)$ and $U(q_1)$ when $\sum p_2 q_2 < \sum p_2 q_1$ and $\sum p_1 q_1 < \sum p_1 q_2$, the quantities would have to be weighted by average utilities which are unknown. Due to diminishing marginal utility, the magnitude of average utility is greater than that of marginal utility. The former exceeds the latter by a greater amount, the less the elasticity of demand. ①

2.5 The Ordinal Nature of Utility and the Bias of Quantity Indexes

Though we have proved that, if $Pq=1$, $Tq>1$, or what is the same thing, $Tq>Pq$ when $Pq=1$, we should not jump to the conclusion that $Tq>Pq$ without qualification. Suppose $Pq=2$, how can we say that $Tq>2$? The indifference map shows that one indifference curve is higher than another, but not how much higher because we cannot measure utility as a numerical quantity. ②While we know that spending $2 yields more utility than spending $1, we cannot say how much more. "The 'true' quantity index, even if successfully obtained, in general has still no definite numerical meaning. It is a magnitude defined solely in its relation to unity." ③

---

① Pigou A C. The Veil of Money[M]. London: MacMillan, 1949: 60-61.

② Since utility is a non-measurable magnitude, Pareto has introduced the term "index of utility," which may be expressed as a function of the utility function, $U=f(u)$, to indicate merely the ordering of preference. Therefore, it is quite immaterial whether the particular form of utility function is $u=\sum pq$, $u=\sqrt{\sum pq}$ or $u=\log \sum pq$ (the last known as Bernouilli's Law), or something else. See Allen R G D. Mathematical Analysis for the Economists[M]. London: MacMillan, 1938: 290-1, 314.

③ Leontief W. Composite Commodities and the Problem of Index Numbers[J]. Econometrica, 1936, 4: 50.

If it can be argued that the marginal utility of money is diminishing,[1] then the greater the magnitude by which Pq exceeds unity, the less likely that $Tq>Pq$. If Pq is far greater than unity, it is possible that $Tq<Pq$. For the same reason, we can say only that $Lq>Tq$ when $Lq=1$. If the argument for diminishing marginal utility of money is valid, it is possible that $Lq<Tq$ when Lq assumes a value far less than unity. To say that Lq has an upward bias and Pq has a downward bias, i. e., $Lq>Tq>Pq$, implies the assumption that the marginal utility of money is constant and hence utility varies in the same proportion as expenditure.

Since $Lq>Tq$ when $Lq=1$ and $Tq>Pq$ when $Pq=1$, we are able to say $Lq>Tq>Pq$ when $Lq=1=Pq$. But, it is then contradictory to have $Tq>1$ and $Tq<1$ at the same time. Therefore, except in the trivial case in which neither prices nor quantities have changed from Situation Ⅰ to Situation Ⅱ, $Lq=1=Pq$ cannot happen.

If $Pq=1$, $U(q_2)>U(q_1)$; by introducing $\bar{q}_2$ as a hypothetical bundle of goods which would have been bought at p_1 prices such that $U(\bar{q}_2)=U(q_2)$, we have $U(\bar{q}_2)>U(q_1)$. Hence $\sum p_1\bar{q}_2>\sum p_1q_1$. As the amount of expenditure represented by $\sum p_1\bar{q}_2$ for the amount of utility obtained at p_1 prices is a minimum, $\sum p_1q_2>\sum p_1\bar{q}_2$. Therefore, $\sum p_1q_2>\sum p_1q_1$ or $Lq>1$. The relation that $Pq=1$ entails $Lq>1$ can also be seen in Figure 1 from the fact that a line drawn through any point (designated as q_2) on the offer curve ff' parallel to the expenditure line ab must lie outside that expenditure line. By analogous processes, it can be shown that $Lq=1$ entails $Pq<1$.

As will be shown, even $Lq>Pq$ does not always hold true.[2] Caution

---

[1] Both Fisher and Frisch attempt to measure the marginal utility of money. While Fisher finds that the curve of diminishing marginal utility of money is steeper than a rectangular hyperbola, Frisch finds that it is flatter. Both, in their methods of derivation, make the implicit assumption that there can be found some commodity with an independent utility. See Fisher I. A Statistical Method for Measuring "Marginal Utility" and Testing the Justice of a Progressive Income Tax[M]//Economic Essays Contributed in Honor of John Bates Clark. New York: MacMillan, 1927: 185; 193; Frisch R. New Methods of Measuring Marginal utility[M]. Tubingen: Mohr, 1932: 64.

[2] See section 7 of this chapter.

must be used in labeling Pq as downward biased and Lq as upward biased.

2.6 Relationship Between Price and Quantity Indexes

An indirect way to obtain the quantity index would be to divide the value index by the price index. If the price index is a "true" one, the quantity index so obtained should be also "true". As a true price index gives an exact (numerical) measure of the change in the price level for obtaining a specific amount of utility, does it follow that the quantity index so derived also gives an exact measure of the change in utility?

In general, corresponding to every given amount of utility, there is a different true price index. Thus, if $U(q_1) \neq U(q_2)$, there will be two true price indexes between Situation Ⅰ and Situation Ⅱ; $Tp(U_1)$ and $Tp(U_2)$.① Consequently, by dividing the value index by the true price indexes, we obtain two quantity indexes. But there is only one true quantity index because each bundle of goods yields a unique amount of utility.② It is, therefore, unsatisfactory to say that the value index divided by the true price index gives a numerically exact measure of the change in utility.

However, if the numerical meaning of the quantity indexes is disregarded and they are interpreted solely in relation to unity, the value index deflated by any true price index will give the true quantity index. For, if $U(q_2) > U(q_1)$, both $V/Tq(U_1)$ and $V/Tp(U_2)$ must be greater than unity.

The above argument suggests the usefulness of true price indexes as a means of deriving true quantity indexes. Let us examine first how the Laspeyres' and the Paasche's price indexes differ from the true price

---

① Except under the assumption of proportionality of expenditure as described by Frisch in his Op. cit. , pp. 13 and 25. See also, Haberler G. Der Sinn Der Indexzahlen[M]. Tubingen: Mohr, 1927: 89 et seq.

② Our position differs from the practice that the "true" quantity index is numerically measurable by the relative amounts of expenditures for different levels of real income at any given set of relative prices. See Rothbarth E. The Measurement of Changes in Real Income under Conditions of Rationing [J]. Review of Economic Studies, 1940, 8(1): 103; Allen R G D. The Economics of Index Numbers[J]. Economica(New Series), 1949, 16(63): 199.

indexes. In the Laspeyres' price index, $\sum p_2 q_1 / \sum p_1 q_1$, $\sum p_2 q_1$ represents an amount of expenditure more than necessary to obtain the amount of utility equivalent to $U(q_1)$. By defining $\bar{q}_1$ as the bundle of goods which would have been bought at p_2 prices such that $U(\bar{q}_1) = U(q_1)$, we have $\sum p_2 q_1 > \sum p_2 \bar{q}_1$. Since $Tp(U_1) = \sum p_2 \bar{q}_1 / \sum p_1 q_1$, we know that $Lp > Tp(U_1)$. By similar reasoning, we know $Tq(U_2) > Pp$. ① Thus it can be said that Lp is the upper limit for $Tp(U_1)$ and Pp is the lower limit for $Tp(U_2)$.

In order to delimit the true price indexes, we have to find the lower limit for $Tp(U_1)$ and the upper limit for $Tp(U_2)$. Pp cannot be the lower limit for $Tp(U_1)$ and Lp cannot be the upper limit for $Tp(U_2)$ unless $U(q_1) = U(q_2)$. ② But, if $U(q_1) = U(q_2)$, we would have no need of any limit and the true price index is simply given by V. If $U(q_1) \neq U(q_2)$, it is not necessarily true that Lp and Pp serve as double limits. Only when $Tp(U_1) \geq Tp(U_2)$, will Lp be an upper limit and at the same time Pp be a lower limit for both $Tp(U_1)$ and $Tp(U_2)$. If $Tp(U_1) < Tp(U_2)$, it is possible that $Tp(U_2)$ exceeds Lp and $Tp(U_1)$ falls below Pp.

If it happens that $\sum p_1 q_2 = \sum p_1 q_1$, we know that $U(q_1) > U(q_2)$. Then, $\sum p_2 q_2 < \sum p_2 \bar{q}_1$. Dividing both sides of the inequality by the above equality, we have

$$\frac{\sum p_2 q_2}{\sum p_1 q_2} < \frac{\sum p_2 \bar{q}_1}{\sum p_1 q_1}, \text{ i. e. } Pp < Tp(U_1).$$

Analogously, it can be shown that if $\sum p_2 q_2 = \sum p_2 q_1$, we have

$$\frac{\sum p_2 q_1}{\sum p_1 q_1} > \frac{\sum p_2 q_2}{\sum p_1 \bar{q}_2}, \text{ i. e. } Lp > Tp(U_2).$$

---

① For a graphical proof, see, among others, Schultz Henry. A Misunderstanding in Index Number Theory[J]. Econometrica, Vol. 7, 1939, 7(1):4; Ulmer M J. The Economic Theory of the Cost of Living Index Numbers[M]. New York: Columbia University Press, 1949:32-39.

For a general mathematical proof, see Hicks J R. Consumers' Surplus and Index Numbers[J]. Review of Economic Studies, 1941, 9:134. where Hicks terms $Tp(U_1)$ as the compensation index and $Tp(U_2)$ as the equivalence Index.

② If $U(q_1) = U(q_2)$, $Tp(U_1) = Tp(U_2)$. That Lp and Pp serve as double limits for the true price index when $U(q_1) = U(q_2)$ has been proved by Keynes J M. A Treatise on Money[M]. New York: Harcourt, 1930:110. Allen R G D. On the Marginal Utility of Money and Its Application[J]. Economica, 1933, 13(5):204.

Thus, in this situation we have shown the lower limit for $Tp(U_1)$ and the upper limit for $Tp(U_2)$. ① It is probably unreasonable to expect either $\sum p_1 q_2 = \sum p_1 q_1$ or $\sum p_2 q_2 = \sum p_2 q_1$ for any particular individual. However, q_1 and q_2 here need not be quantities for the same individual. As long as they are the quantities of any two individuals who have the same tastes, the limits derived above are valid. It is hoped that family budget data might provide a check on the validity of these conditions, i. e. $\sum p_1 q_2 = \sum p_1 q_1$ and $\sum p_2 q_2 = \sum p_2 q_1$. ②

The narrower the limits within which the true price indexes lie the better. Unless the limits which we ascertain are between Lp and Pp, they would not help in reducing the region of ignorance in the comparison of utility. For, if Lp is known to be the upper limit, $V/Lp = Pq$ gives the lower limit for Tq in its relation to unity; if Pp is known to be the lower limit, $V/Pp = Lq$ gives the upper limit for Tq in its relation to unity. Here, we come to the same conclusion as before and derive nothing new from the limits for the true price indexes. ③

2.7 The Meaning of the Engel Curve

When relative prices remain unchanged, the changes in the quantities of the commodities purchased as a result of the change in real income forms an expenditure path technically known as the Engel curve. Let us define marginal proportion of expenditure for a commodity as the proportion of an

---

① These are the techniques described in Staehle Hans. A Development of the Economic Theory of Price Index Numbers[J]. Review of Economic Studies, 1935, 2(3):180.

② For a practical appraisal, see Allen R G D. Some Observations on the Theory and Practice of Price Index Numbers[J]. Review of Economic Studies, 1935, 3(10):57-66.

Note that if q_1 and q_2 are not the quantities of the same individual, the lower limit for $Tp(U_1)$, which we find, is not the same as that for $Tp(U_2)$ and the upper limit for $Tp(U_2)$, which we find, is not the same as that for $Tp(U_1)$.

③ For example, A. P. Lerner gives the highest price ratio $(p_2/p_1)_h$ between the two situations as the upper limit for $Tp(U_2)$ and the lowest price ratio $(p_2/p_1)_l$ as the lower limit for $Tp(U_1)$. Thus we have

$(p_2/p_1)_h > Tp(U_2) > Pp$ and $(p_2/p_1)_l < Tp(U_1) < Lp$

It is obvious that the upper limit for $Tp(U_2)$ is set too high and the lower limit for $Tp(U_1)$ too low. See idem, A Note on the Theory of Price Index Numbers[J]. Review of Economic Studies, 1935, 3(10):55.

additional dollar of expenditure devoted to that commodity. If fixed proportions of the total expenditure are allocated among different commodities regardless of the change in the level of income (i. e. the income elasticity of demand for each commodity is unity when all money income is spent), we have a case of constant marginal proportion of expenditure.

It has been shown that $Lp>Tp>Pp$ when $U(q_1)=U(q_2)$. If $U(q_1) \neq U(q_2)$, we can, however, put $U(q_1)=U(kq_2)$ where $kq_2=k'q_2'+k''q_2''+\cdots +k^s q_2^s$ representing the actual quantities of the s commodities which would have been bought at p_2 prices for an amount of real income equivalent to $U(q_1)$. If $k'=k''=\cdots=k^s=K$, as in the case of constant marginal proportion of expenditure, we have $U(q_1)=U(Kq_2)$. The expenditure for $U(Kq_2)$ is accordingly $K\sum p_2 q_2$ at p_2 prices and $K\sum p_1 q_2$ at p_1 prices. Since we put $U(q_1)=U(Kq_2)$, $\sum p_2 q_1 > K\sum p_2 q_2$ and $\sum p_1 q_1 < K\sum p_1 q_2$ and hence

$$\frac{\sum p_2 q_1}{\sum p_1 q_1} > \frac{K\sum p_2 q_2}{K\sum p_1 q_2}.$$

The right side of this inequality tells us that the value of Pp would remain the same at all levels of real income, therefore, we have not only $Lp>Pp$, but also $Lp>Tp>Pp$.

In real life, however, we do not have a uniform constant marginal proportion of expenditure especially when there is a substantial change in real income.[①] Let us consider the substitution effect of a change in relative prices together with a change in real income on the purchases of two groups of commodities; one representing necessities and the other representing luxuries.

In Figure 3, the quantity of luxuries is measured along OX and the quantity of necessities is measured along OY. Situations Ⅰ and Ⅱ are

---

① The Engel's law reads: "As income increases, the expenditures on different items of the budget have changing proportions and that the proportions devoted to the more urgent needs (such as food) decrease while those devoted to luxuries and semi-luxuries increase." This law has been substantiated by Allen and Bowley's findings. See Allen R G D, Bowley A L. Family Expenditure[M]. London: King, 1935: 7.

indicated by B and G respectively while ii' is the expenditure line at B and jj' the expenditure line at G. Draw through B hh' parallel to jj' and through G kk' parallel to ii'. Oh, Oi, Oj and Ok (also, Oh', Oi', Oj' and Ok') then correspond to the expenditures $\sum p_2 q_1, \sum p_1 q_1, \sum p_2 q_2$ and $\sum p_1 q_2$ respectively. A straight line OF, indicating constant marginal proportion of expenditure, is drawn from the origin O through B, which intersects jj' at A. Through A, we draw cd parallel to kk'. OBE is drawn to represent the Engle curve. Since $Oh/Oi = Oj/Od, Oh/Oi < Oj/Ok$, that is, $\sum p_2 q_1 / \sum p_1 q_1 < \sum p_2 q_2 / \sum p_1 q_2$. Here we have a case of $Pp > Lp$. ①

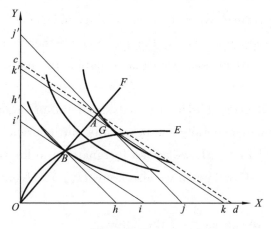

Figure 3　Luxuries

From the Engel curve in Figure 3 we see that, if the Necessities price ratio between luxuries and necessities remains the same, the greater the increase of real income from Situation Ⅰ to Situation Ⅱ, the farther the point G lies to the right side of OF. Then a change in relative prices in favor of luxuries will push the point G still farther to the right while a change in relative prices in favor of necessities will bring it closer to OF. Suppose, as in Figure 3, the price ratio changes in favor of necessities, the point G will lie to the right or left side of OF according as the substitution effect of the change in relative prices is less or greater than the income effect on the proportions of luxuries and necessities purchased.

---

①　This case has been considered by Hicks in his, *Op. cit.*, p. 135 and A. L. Bowley in his article "Earnings and Prices, 1904, 1914, 1937-8," Review of Economic Studies, 1940, 8(1):135.

Under the condition that G represents a higher level of income than B and the price ratio changes in favor of necessities, we will have $Pp > Lp$ as long as G falls within XOF. Reversing the situations so that real income falls from G to B and at the same time the price ratio changes in favor of luxuries, we also have $Pp > Lp$ as B falls within YOG (these arguments can be easily verified from the relation: $Oh/Oi < Oj/Ok$). ①

From the identical relation, $V = Pq \cdot Lp = Lq \cdot Pp$, we know, if $Pp > Lp, Pq > Lq$.

2.8 The Index Number Criterion for Increase of Real Income

Due to the ordinal nature of true quantity indexes, the comparison of real income consists of answering the question whether there is an increase (or decrease) of real income from one situation to another and not how much increase.

We have not been able, in the analysis of the price and the quantity indexes, to eliminate generally and practically our ignorance of the cases in which $Lq > 1$ and $Pq < 1$. If we are so fortunate as to know that $U(q_2) > U(q_3)$ and $U(q_3) > U(q_1)$, then we also know $U(q_2) > U(q_1)$ even if $\sum p_2 q_2 < \sum p_2 q_1$. ②

In general, we can only say the following:

(1) If $Pq \geq 1$, real income is at least higher in Situation Ⅱ than in Situation Ⅰ.

(2) If $Lq \leq 1$, real income is at least lower in Situation Ⅱ than in Situation Ⅰ.

3. The Valuation of Real Income for a Group of Individuals

The comparison of aggregate real income for a group of individuals leads to considerations of income distribution. If income distributions are different in different situations, the valuation of real income may involve

---

① If the change in real income from B to G is conditioned by $Pq = 1$, G will be a point on Bh', i.e. within YOF or outside XOF, and, so, $Lq > Pq$.

② This may be the case when the difference in relative prices between p_1 and p_2 is great and the differences between p_1 and p_3 and between p_2 and p_3 are small. Cf, Samuelson, op. cit., p. 152.

interpersonal comparisons of utility. In such as an objective standard for comparing different individuals' utility is lacking, difficulties in making real national income comparisons are apparent. This is especially true when knowledge of what has happened is restricted to aggregate quantity index numbers alone——since the latter are ordinarily constructed without recognition of changes in the distribution of income. The difficulties suggested have led to a divergency of propositions as to a meaningful definition of increases in real national income. In this chapter the meaning and use of aggregate quantity index numbers will be examined according to various definitions of real national income.

3.1 The Problem of Distribution

For simplicity, consider a group consisting of two persons A and B. If both A and B's real income have increased from one situation to another, we certainly would say that the group's real income has increased. If both A and B's real incomes have decreased, we then say that the group's real income has decreased. But what can we say if A's real income has increased while B's has decreased, or A's has decreased while B's has increased? In this case, we may be inclined to say that it depends on the respective magnitudes of the increase and the decrease. Yet, no matter how different are these two magnitudes, to say the group's real income has increased or decreased involves a dubious interpersonal comparison of utility. For example, granted the possibility of measuring the individual's real income precisely, if A's real income has been doubled while B's has been reduced to nine-tenth, a simple arithmetic mean of the ratios yields a value far greater than unity. In this sense then, there has been an increase of the group's real income. But, what would prevent us assigning different weights to these two ratios? For example, weights corresponding to expenditures, social values of individual utilities, capacities of feeling satisfaction, etc?

How can we tell that a dollar placed with one person would be worth more, as much as, or less than, with the other? Given an equilibrium situation where goods or services cannot be further exchanged to mutual benefit, a redistribution of these goods and services will merely make one person better off and the other worse off. Is it possible to say that the situation before the redistribution is better or worse for the group as a whole than after it? If we cannot appraise distributions, can we ever avoid doing so and still make real income comparison for the group in any sense?

3.2　Historical Background

A review of the literature of the subject may help lay down the plan for the present analysis. Let us note, without elaboration, how the problem of distribution has been treated in the valuation of real national income in the recent past. To go back no further than to Pigou's Economics of Welfare we find the following in his first(1920) edition:

> I define a change in the dividend, or real income, of a group as follows:.... this quantity is larger or smaller in period A than in period B, according as the amount of economic satisfaction derived from it by a representative member of the group——insofar as the tastes and temperament of the group have not been changed by external causes (i. e. otherwise than as an indirect effect of their changed purchases)——is larger or smaller. [1]

Thus, the problem of treating distribution is reduced to finding the representative member of the group. But how to find the representative member is not shown.

In 1932, in his fourth edition of the same work, he changed this definition to say:

> From the point of view of period I, an increase in the size of the dividend for a group of given numbers is a change in its content such that, if tastes in period II were the same as those prevailing in period I and if the distribution of purchasing power

---

[1]　Pigou A C. The Veil of Money[M]. London: MacMillan, 1949: 70.

were also the same as prevailed in period Ⅰ, the economic satisfaction (as measured in money) due to the items added in period Ⅱ is greater than the economic satisfaction (as measured in money) due to the items taken away in period Ⅱ.... From the point of view of period Ⅱ, an increase in the dividend is defined in exactly analogous ways. ①

Then, he concludes:

> Since tastes and distribution are unaltered,... if both our fractions (Lq and Pq) are greater than unity, it necessarily follows that the economic satisfaction yielded by the collection q_2 bought in the second period is greater than the economic satisfaction yielded by the collection q_1 bought in the first period. By analogous reasoning it can be shown that, if both the above fractions are less than unity, the converse result holds good. ②

In fact, however, a change in the national income (such as resulting from the removal of a tariff or the enactment of the unemployment legislation) is usually accompanied by a change in the distribution of income. The assumption of constancy of distribution merely shows our inability to make any positive statement for comparing distributions and, consequently, leaves real national income comparison on a shaky ground.

In 1939, Kaldor enunciated the principle of compensation saying that if after an economic change some gain while others lose, real national income

---

① Pigou A C. The Veil of Money[M]. London: MacMillan, 1949: 54.
② Pigou A C. The Veil of Money[M]. London: MacMillan, 1949: 63.

is said to increase if the gainers can over compensate the losers. ① This principle was later blessed by Hicks as though furnishing solid ground for making comparisons of real national income by taking proper care of changes in distribution yet without making any value judgment concerning distribution. ②

In 1940, Hicks gave an important new meaning to the aggregate index number relation $\sum p_2 q_2 > \sum p_2 q_1$. He said:

> ... What does it signify if $\sum p_2 q_2 > \sum p_2 q_1$?
>
> ... since this condition refers only to the total quantities acquired, it can tell us nothing about the distribution of wealth among the members of the group. There may be a drastic redistribution of wealth among the members and the aggregates will remain exactly the same. Thus what the condition $\sum p_2 q_2 > \sum p_2 q_1$ tell us is that there is some distribution of the $q_1 s$ which would make every member of the group less well off than he actually is in the II situation. For if the corresponding inequality were to hold for every individual separately, it would hold for the group as a whole.
>
> As compared with this particular distribution, every other

---

① Kaldor Nicholas. Welfare Propositions of Economics and Interpersonal Comparisons of Utility [J]. Economica,1939,49(9):549-552.

Prior to this article, the idea of the principle of compensation, however, had been conceived by Viner J. Studies in the Theory of International Trade[M]. New York: Harper,1937:533-534. Hotelling H. The General Welfare in Relation to Problems of Taxation and of Railway and Utility Rates [J]. Econometrica,1938,6(7):258; Samuelson. The Gain from International Trade[J]. Canadian Journal of Economics and Political Science, 1939. Reprinted in Readings in the Theory or International Trade, Philadelphia: Blakiston, 1949: 251. But it is Kaldor who advocates wholeheartedly this principle as providing a scientific basis for welfare economics.

Kaldor's article was a formal challenge to the views of Harrod's "Scope and Method of Economics," Economica,1938,48(9):396-397, and Robbins. Interpersonal Comparisons of Utility: A Comment[J]. Economica,1938,48(12):638, by showing the possibility of making welfare propositions independent of interpersonal comparison.

② Hicks J R. The Foundations of Welfare Economics[J]. Economica,1939(12):698. Idem. The Rehabilitation of Consumer's Surplus[J]. Review of Economic Studies,1940,8(1):111.

distribution of the q'_1s would make some people better off and some worse off. Consequently if there is one distribution of the q'_1s in which every member of the group is worse off than he actually is in the Ⅱ situation, there can be no distribution in which every one is better off, or even as well off...

...Let us say that the real income of society is higher in Situation Ⅱ than in Situation Ⅰ, if it is impossible to make every one as well off as he is in Situation Ⅱ by any redistribution of the actual quantities acquired in Situation Ⅰ. [1]

In stating this supposedly <u>wert-frei</u> definition of increase of real national income, Hicks erred in attaching a trical importance to the two situations compared. A correction was made a year later by Scitovszky who added another condition to the Hicks definition: namely, it also must be "possible in the new situation so to redistribute income as to make everybody better off than he was in the initial situation." [2] Later, Baumol and Samuelson expressed desire for a definition of increase of real national income such that Situation Ⅱ is said to be better than Situation Ⅰ only if in all possible distributions (actual or potential) q_2 would make everyone better off than q_1. Only such a definition is free from the charge of any undue sanctification of the status quo. [3]

It must be realized that it is the <u>potential</u> rather than the <u>actual</u> real national income which the compensationists want to compare. They ask whether there is an increase or decrease of real national income if the $q's$ are distributed in such and such a fashion. The compensationists cannot tell whether one situation is actually better than another. Only when the potential redistributions are accomplished (i. e. the compensations are made) will the situation which is potentially better become actually better. There

---

[1] Hicks. The Valuation of the Social Income[J]. Economics,1940 (1):110-111.

[2] Scitovszky T. A Note on Welfare Propositions in Economics[J]. Review of Economic Studies, 1941,9(November):86.

[3] Baumol W J. Community Indifference[J]. Review of Economic Studies, 1946—1947,14:46. Samuelson. Evaluation of Real National Income[J]. Oxford Economic Papers,1950,2(1):10.

fore, Arrow remarks, unaccomplished distributions are irrelevant and the principle of compensation involves the value judgments inconsistent with "the possibility of rational choice by the community as a whole." ①

To interpret changes in actual real national income, except in the very rare case in which all individuals' real income changes in the same direction, requires some scheme of interpersonal comparisons. ②This is, however, beyond the scope of the present study.

3.3 Definitions of Real National Income

When we are concerned with a single individual, we can define real income unambiguously as the amount of utility or satisfaction which is derived from the commodities which he receives or purchases. But when we consider a <u>group</u> of people or a community, we can no longer, without ambiguity, define real income in a similar way. For, given the national product, the total amount of utility (or national economic welfare) derived from it depends on its distribution. ③

From the preceding section we may see the possible divergency of definitions of real national income. First, by ignoring the distribution of income or by assuming its distribution to be fixed, real national income may be defined simply as the amount of utility derived from the national product as if it were consumed by a single individual. Second, according to the compensation principle, real national income may be defined as the total amount of utility which would be derived from the national product when distributed in one, two, or any number of ways. Third, if it is held that only the actually accomplished distributions are relevant, real national income may be defined as the total amount of utility derived from the national product as it is actually distributed.

---

① Arrow K J. Social Choice and Individual Values, Cowles Commission[M]. New York: Wiley, 1950:45.

② For exposition of this problem, reference may be made to Lange O. The Foundations of Welfare Economics[J]. Econometrica, 1942, 10(3—4):215-224.

③ By saying "The total amount of utility," we mean neither that utility is measurable, nor that different individuals' utilities are objectively comparable. All we purport to do is to convey the aggregate concept of real national income in one way or the other.

Each definition should be appraised by two standards: namely, how logical it is and how adaptable it is to measurement. In what follows, we shall investigate specifically (1) what aggregate quantity indexes can tell us about changes in real national income and (2) the logic of accepting these changes as a criterion for determining an increase or decrease in the real national income from one situation to another.

3.4 Statement of Assumptions

In addition to the assumptions set forth in the last chapter,① the following assumptions are pertinent:

(1) The population consists of the same individuals in the situations compared.

(2) All goods are consumption goods consumed by individuals.

(3) Real national income is an increasing function of all the individuals' utilities and each individual's utility depends solely on the goods he receives.

3.5 Individual vs. Aggregate Index Numbers

In the following analysis, the notations used in the last chapter for the individual will also be used for the nation as a whole. While we make no difference between the prices for the individual and those for the nation, we will distinguish the individual's quantities with a superscript. For example, the Paasche's individual quantity index will now be designated by $\sum p_2 q_2^* / \sum p_2 q_1^*$ or $P^* q$, whereas $\sum p_2 q_2 / \sum p_2 q_1$ or Pq will denote the aggregate quantity index.

For an individual consumer, we say that $\sum p_2 q_2^* > \sum p_2 q_1^*$ means $U(q_2^*) > U(q_1^*)$ because he chooses to buy q_2^* rather than q_1^* (q_1^* cost him less than q_2^*). For the nation as a whole, such an interpretation may not be valid.

A change in one individual's demand can hardly influence market prices, but a change in the nation's demand certainly may. Unless all

---

① They are: (1) constancy of tastes, (2) same variety of goods, (3) optimal allocation of expenditures and (4) utility as an increasing function of expenditures.

supply prices are assumed constant, we cannot say from the aggregate relation $\sum p_2 q_2 > \sum p_2 q_1$ that the nation chooses to buy q_2 rather than q_1, for q_1 may not be bought for the amount of expenditure equivalent to $\sum p_2 q_2$. However, in an attempt to show $U(q_2) > U(q_1)$, — under the assumption that the distribution of purchasing power in Situation I is the same as prevailed in Situation II — Pigou once argued: "In real life, with a large number of commodities, it is reasonable to suppose that the upward price movements caused by shifts of consumption would roughly balance the downward movements."[①] So q_1 can still be bought with the amount of expenditure equivalent to $\sum p_2 q_2$.

In order to show that $U(q_2) > U(q_1)$ from the relation $\sum p_2 q_2 > \sum p_2 q_1$, however, Pigou's argument is essentially unnecessary. Since, at p_2 prices, q_2 is assumed to represent the equilibrium point in which everybody's expenditures are so optimally allocated that the price ratio between any two goods is equal to their marginal rate of substitution, any further substitution of one good for another would mean a displacement from equilibrium and therefore result in a loss of utility. [②] Here, q_1 can be said to represent such a displacement when $\sum p_2 q_2 = \sum p_2 q_1$ (for, to further substitute one good for another at the same p_2 prices involves no change in expenditure). Therefore, though q_1 may not actually be purchased for the amount of expenditure equivalent to $\sum p_2 q_2$, we are, however, sure that $U(q_2) > U(q_1)$ if $\sum p_2 q_2 = \sum p_2 q_1$. $U(q_2)$ will exceed $U(q_1)$ by an even greater amount if $\sum p_2 q_2 > \sum p_2 q_1$.

3.6 Under Constancy of Distribution

What we have just learned is that, with the distribution of purchasing power as it is in Situation II, it is better for everyone, and hence for the nation as a whole, to have q_2 instead of q_1. But, is it true that everyone is better off with q_2 in Situation II than with q_1 in Situation I ? This is true only when the distribution of purchasing power is unchanged.

---

① Pigou A C. Economics of Welfare[M]. 4th ed. London: MacMillan, 1932: 61-62.

② Assuming that there is only one point of tangency between the price line (or plane) and the indifference curve (or surface) for any individual.

Let $\bar{q}_1$ be the bundle of goods which would have been bought at the p_2 prices so as to make everyone just as well off as he was in Situation Ⅰ. That is, $U(\bar{q}_1)=U(q_1)$. Since $\bar{q}_1$ represent an equilibrium point at p_2 prices, $\sum p_2\bar{q}_1$ would be a minimum amount of expenditure for the given amount of utility, $U(\bar{q}_1)$ or $U(q_1)$, therefore $\sum p_2 q_1 > \sum p_2 \bar{q}_1$. Then, if $\sum p_2 q_2 > \sum p_2 q_1$, it follows that $\sum p_2 q_2 > \sum p_2 \bar{q}_1$. Under the assumption of constancy of distribution of purchasing power, the difference in the expenditures $\sum p_2 q_2 - \sum p_2 \bar{q}_1$ would be so distributed as to make everyone better off. Thus, we are able to say that $\sum p_2 q_2 > \sum p_2 q_1$ means an increase of real national income from Situation Ⅰ to Situation Ⅱ: $U(q_2) > U(q_1)$.

Once $U(q_2) > U(q_1)$ is established, it can be shown that $\sum p_1 q_2 > \sum p_1 q_1$ in exactly the same manner as we did for an individual in the last chapter. Let $\bar{q}_2$ be the bundle of goods which would have been bought at p_1 prices so as to make everyone as well off as he is in Situation Ⅱ. That is, $U(\bar{q}_2)=U(q_2)$. Since $U(q_2) > U(q_1)$, it follows that $\sum p_1 \bar{q}_2 > \sum p_1 q_1$. As $\sum p_1 \bar{q}_2$ represent a minimum, we have $\sum p_1 q_2 > \sum p_1 \bar{q}_2$. Therefore, $\sum p_1 q_2 > \sum p_1 q_1$.

By analogous processes, it can be shown, under constancy of distribution of purchasing power, that $\sum p_1 q_1 > \sum p_1 q_2$ means $U(q_1) > U(q_2)$ and also entails $\sum p_2 q_1 > \sum p_2 q_2$.

3.7 The Principle of Compensation and the Meaning of $\sum p_2 q_2 > \sum p_2 q_1$

Waiving the assumption of constancy of distribution, what can we now say about the relation $\sum p_2 q_2 > \sum p_2 q_1$ for the nation? The compensationists say that if, after an economic change, the gainers can overcompensate the losers, there is an increase of real national income. Let the situation change from Situation Ⅰ to Situation Ⅱ, can the gainers overcompensate the losers if $\sum p_2 q_2 > \sum p_2 q_1$?

Instead of redistributing q_2 to show whether or not the gainers can overcompensate the losers, Hicks works the other way around by saying that if $\sum p_2 q_2 > \sum p_2 q_1$, there will be some redistribution of q_1 which makes everybody worse off in Situation Ⅰ (before the change) than he actually is in Situation Ⅱ (after the change). The condition of this possible redistribution is then accepted as a definition of an increase of real national income from Situation Ⅰ to Situation Ⅱ.

Let us examine how this redistribution is achieved.

Suppose a community consists of n individuals, we then have
$$q_1 = q_{11}^* + q_{12}^* + \cdots + q_{1n}^*, \text{ and}$$
$$q_2 = q_{21}^* + q_{22}^* + \cdots + q_{2n}^*,$$
where the q^*'s are individuals' quantities in the actual distributions. If $\sum p_2 q_2 > \sum p_2 q_1$, i.e.
$$\sum p_2 q_{21}^* + \sum p_2 q_{22}^* + \cdots + \sum p_2 q_{2n}^* > \sum p_2 q_{11}^* + \sum p_2 q_{12}^* + \cdots + \sum p_2 q_{2n}^*,$$

It is then possible to redistribute q_1 into[①]
$$\hat{q}_{11}^* + \hat{q}_{12}^* + \cdots + \hat{q}_{1n}^* = q_{11}^* + q_{12}^* + \cdots + q_{1n}^*$$
such that
$$\sum p_2 q_{21}^* > \sum p_2 \hat{q}_{11}^*,$$
$$\sum p_2 q_{22}^* > \sum p_2 \hat{q}_{12}^*,$$
$$\cdots\cdots$$
$$\sum p_2 q_{2n}^* > \sum p_2 \hat{q}_{1n}^*.$$

According to the individual index number criteria, we know
$$U(q_{21}^*) > U(\hat{q}_{11}^*),$$
$$U(q_{22}^*) > U(\hat{q}_{12}^*),$$
$$\cdots\cdots$$
$$U(q_{2n}^*) > U(\hat{q}_{1n}^*).$$

Thus, everybody is made better off with his actual share in Situation Ⅱ than with the potential redistributed share in Situation Ⅰ.

3.8 Requirement of Optimum Equilibrium[②]

As pointed out by Samuelson, there is a step missing in the Hicks' analysis discussed above. If the redistribution arrived at as suggested by this analysis does not reach the equilibrium point of exchange at p_1 prices, people may benefit by trading with each other. May it happen then that someone's position is eventually improved as a result of the redistribution?

---

① Assuming the units of all commodities are small enough.
② The method of analysis in this section is largely Samuelson's. See Samuelson P A. Evaluation of Real National Income[J]. Oxford Economic Paper, 1950, 2(1):8.

In order for the analysis to be valid, it must be shown that the redistribution is accomplished so that no one can gain without imposing a loss on someone else. ①

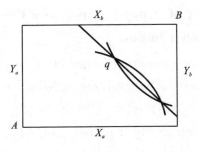

Figure 4

Suppose, for simplicity, that there are only two individuals A and B with two commodities X and Y and that the total quantities in Situation I are defined in the box diagram in Figure 4. A's shares of X and Y are measured by X_a and Y_a. B's shares are measured by X_b and Y_b. Suppose the redistribution represented by the point q fulfills the requirement of $\sum p_2 q_2^* > \sum p_2 \hat{q}_1^*$ for both A and B, but at q neither A nor B is in optimum equilibrium as their respective indifference curves are not tangential to each other at q. Let a straight line be drawn through q to represent the p_2 prices, then any point on this price line will represent an equal amount of expenditure and will, therefore, fulfill the requirement of $\sum p_2 q_2^* > \sum p_2 \hat{q}_1^*$ for both A and B. Since the contract curve (by which we mean the locus of the points of tangency to both A and B's indifference curves) must

---

① This missing step has caused some confusion. For example, Little has managed to refute the Hicks criterion by comparing two situations in which the efficiencies in the allocation of goods are achieved in different degrees. Suppose the allocation of q_2 in Situation II is not optimal, then it may be possible to so redistribute q_1 as to make everybody better off than he actually is in Situation II even if $\sum p_2 q_2 > \sum p_2 q_1$. See Little I M D. A Critique of Welfare Economics[M]. Oxford: Clarendon, 1950: 99-104. While his argument is proper in welfare economics, it has nothing to do with the index number criteria which already presume the prevalence of the condition of optimal allocation of goods.

Scitovszky has specified three cases of real national income comparisons between two situations: (1) the allocations of goods are optimal in both, (2) the allocations are optimal in one and suboptimal in the other and (3) the allocations are suboptimal in both. It is the first case which is relevant to the index number criteria. See Scitovszky, Op. cit., p. 86.

intersect the p_2 price line somewhere,① this point of intersection will be a point of redistribution which leaves everyone in an inferior position (as well as being an optimum equilibrium point).

3.9 Inadequacy of $\sum p_2 q_2 > \sum p_2 q_1$ as a Criterion of Increase of Real National Income

We now proceed to examine the logic of Hicks' definition of increase of real national income in light of the redefined meaning of the index number relation $\sum p_2 q_2 > \sum p_2 q_1$.

Is it the same thing to say that there could be a redistribution of q_1 which makes everyone worse off in Situation I than he actually is in Situation II as to say that there could be a redistribution of q_2 which makes everyone better off in Situation II than he actually was in Situation I?

In order to tell which is better between q_1 and q_2 in terms of real national income, the comparison of a possible redistribution of q_1 with the actual distribution of q_2 implies the propriety of the actual distribution in

---

① But, assuming these two individuals are very different in tastes and the two commodities under consideration are hardly substitutable for either individual, it is possible that the contract curve does not intersect the price line in the section confined by the box (see Figure 5). In this case, the equilibrium of exchange cannot be achieved by the required redistribution. However, the intercept, q', by the price line on the border line will be a point in Which no one can gain without imposing a loss on the other except in the very rare case in which each individual consumes a different commodity exclusively. In this latter case, a point either on the upper left or the lower right corner represents the best point of distribution.

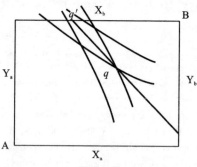

Figure 5

In view of the large number of consumers, each of which may trade with the other, and the large number of substitutes which an individual may consume, it is most likely that an equilibrium price will be concluded for each commodity after the redistribution of purchasing power.

Situation II (symbolized as D_2). On the other hand, the comparison of a possible redistribution of q_2 with the actual distribution of q_1 implies the propriety of the actual distribution in Situation I (symbolized as D_1). Scitovszky, to avoid discriminating between the two actual distributions, would make both comparisons. The Scitovszky condition for the increase of real national income from Situation I to Situation II requires that $U(q_2) > U(q_1)$ at both D_1 and D_2.

Does the criterion $\sum p_2 q_2 > \sum p_2 q_1$ necessitate some redistribution of q_2 so as to make everyone better off in Situation II than he actually was in Situation I, that is, is $U(q_2) > U(q_1)$ at D_1?

Suppose we break down again the aggregate quantities into individuals' quantities. Then the relation

$$\sum p_2 q_2 > \sum p_2 q_1$$

becomes

$$\sum p_2 q_{21}^* + \sum p_2 q_{22}^* + \cdots + \sum p_2 q_{2n}^* > \sum p_2 q_{11}^* + \sum p_2 q_{12}^* + \cdots + \sum p_2 q_{1n}^*.$$

But, in this instance, we must redistribute q_2 into

$$\hat{q}_{21}^* + \hat{q}_{22}^* + \cdots + \hat{q}_{2n}^* = q_{21}^* + q_{22}^* + \cdots + q_{2n}^*$$

such that

$$\sum p_2 \hat{q}_{21}^* > \sum p_2 q_{11}^*,$$
$$\sum p_2 \hat{q}_{22}^* > \sum p_2 q_{12}^*,$$
$$\cdots\cdots$$
$$\sum p_2 \hat{q}_{2n}^* > \sum p_2 q_{1n}^*.$$

if the possibility of making everyone better off is to be demonstrated.

With the prices and the quantities as they stand in the relation $\sum p_2 q_2 > \sum p_2 q_1$, the above processes of separation are correct from the mathematical point of view. However, what would guarantee that the p_2 prices do not change as a result of the redistribution of q_2 in Situation II? This question would not arise if we redistribute q_1 in Situation I since the p_1 prices are not involved in the above index numbers. This question may be put otherwise. If the p_2 prices do not change,[①] then what guarantee is there that after the redistribution everyone is in equilibrium position?

---

① Assuming, for the moment, the existence of strict price controls.

If it is assumed that everyone's indifference map is the same, and that the marginal proportion of expenditure is constant for every commodity over the relevant range of changes in expenditure(implying linearity of the Engel curves), a change in income distribution will not affect the proportions of various goods purchased by the nation as a whole. That is to say, at p_2 prices, consumers will all be in equilibrium with respect to supply and demand with the total purchase of q_2 both before and after the redistribution of purchasing power.

As tastes are dissimilar among individuals and as constant marginal proportion of expenditure(even over the relevant range only) is at best an approximation, a redistribution of purchasing power will be accompanied by a change in relative prices to attain the competitive equilibrium. It is invalid from the economic point of view to separate the aggregate index number on the basis of a redistribution of q_2 without considering the accompanied change in p_2 prices.

3.10 Kuznets' Numerical Example

Showing that $\sum p_2 q_2 > \sum p_2 q_1$ while specifying $U(q_2) > U(q_1)$ at D_2, does not ensure $U(q_2) > U(q_1)$ at D_1, Kuznets, for example, gives the following numerical example in which p and q refer to prices and quantities of necessities and P and Q to those of luxuries.[①]

Situation I

Purchases by	Quantities		Prices		Money Aggregates	
	q	Q	p	P	pq	PQ
Poor	8	0	1	1	8	0
Rich	1	3	1	1	1	3
Total	9	3			9	3

---

① Kuznets Simon. On the Valuation of Social Income—Reflections on Prof. Hicks' Article, Part I [J]. Economica, 1948, 15(Feb): 3.

Situation II

Purchases by	Quantities		Prices		Money Aggregates	
	q	Q	p	P	pq	PQ
Poor	6	0	1	1	6	0
Rich	1	7	1	1	1	7
Total	7	7			7	7

From these hypothetical data, it is easy to show that for <u>both</u> poor and rich $Pq=14/12=1.17$; for the poor alone $Pq=6/8=0.75$ and for the rich alone $Pq=8/4=2$. The change from Situation I to Situation II has made the rich better off and the poor worse off.

Whether or not it is possible to make both rich and poor better off by a shift of purchasing power depends on the degree to which Q can be substituted for q. If Q and q are perfect substitutes, the relative prices would not change as a result of redistribution of purchasing power. In the absence of perfect substitutability, a shift of purchasing power from the rich to the poor will cause P to fall and p to rise.

Using Little's illustration (shown below),[①] a redistribution of purchasing power will result in a rise in p by 25 percent while P falls by 25 percent.

Purchased by	Quantities		Prices		Money Aggregates	
	q	Q	p	P	pq	PQ
Poor	6	4	1.25	0.75	$7\frac{1}{2}+3=10\frac{1}{2}$	
Rich	1	3			$1\frac{1}{4}+2\frac{1}{4}=3\frac{1}{2}$	
					14	

Then, for the poor $Pq=10\frac{1}{2}/10>1$ and for the rich $Pq=3\frac{1}{2}/3\frac{1}{2}=1$. The poor are better off and the rich no worse off. The Scitovszky condition is fulfilled.

Suppose, again, q and Q are less substitutable and that there is a 50

---

① Little I M D. The Valuation of the Social Income[J]. Economica, 1949, 16(61): 14.

percent change in prices, then we have:

Purchased by	Quantities		Prices		Money Aggregates	
	q	Q	p	P	pq	PQ
Poor	6	4	1.50	0.50	\multicolumn{2}{c}{$9+2=11$}
Rich	1	3			\multicolumn{2}{c}{$1\frac{1}{2}+1\frac{1}{2}=3$}
					\multicolumn{2}{c}{$\overline{14}$}

Now, for the poor, $Pq=11/12<1$, and for the rich, $Pq=3/3=1$. While the rich are no worse off, we do not know whether or not the poor are better off. ①

This is only a particular example. The Scitovszky condition can be better understood from a more general approach.

3.11 Community Indifference

For an individual, as far as he is consistent, his indifference curves will not intersect. This is true for community (collective) indifference curves, however, only on the assumption that distribution is constant. An examination of the community indifference curve will make this apparent.

The community indifference curve involves the aggregation of individual indifference curves. For each given set of relative prices we read from each individual's indifference curve a different bundle of goods. Summing these bundles of goods for different individuals (each summation referring to a different set of relative prices), we arrive at the community indifference curve. If all the individual indifference curves are continuous and convex to the origin, the community indifference curve will be also continuous and convex to the origin.

Under the assumption of constancy of distribution, any point above a given community indifference curve will make everyone better off and any point below it will make everyone worse off.

---

① It may be pointed out that both Kuznets and Little's figures do not live up to the presumption of diminishing marginal rate of substitution, since in Kuznets' figure, different quantities are purchased at the same relative prices whereas in Little's figure, the same quantities are purchased at different relative prices. Therefore, $Pq=1$ here does not of necessity imply $U(q_2)>U(q_1)$.

Therefore, the community indifference curve which goes through any point above (or below) a given community indifference curve will lie uniformly above (or below) that given community indifference curve. In other words, they do not intersect each other.

However, the community indifference curves may not necessarily be separate from each other if distribution is allowed to change. Due to the difference in tastes among individuals and the changing marginal proportion of expenditure, through any given point (representing a given bundle of goods) there can be drawn as many community indifference curves as the number of possible distributions. ①

An example will illustrate the point. Suppose there are only two individuals A and B with two commodities X(say, wine) and Y(say, bread). Let us assume that A has a greater preference for wine than B and that the proportion of expenditure is in favor of wine as real income increases. By increasing the price ratio of Y to X, A and B's respective indifference schedules are shown below:

Distribution I

	Commodities			Quantities				
A	X_a	4.5	$\underline{5}$	6	7.5	9.5	15
	Y_a	6	$\underline{5}$	4	3	2	1
B	X_b	3.5	$\underline{5}$	7	9.5	15	20
	Y_b	6	$\underline{5}$	4	3	2	1
Community	X	8	$\underline{10}$	13	17	24.5	35
	Y	12	$\underline{10}$	8	6	4	2

Distribution II

	Commodities			Quantities				
A	X_a	6.5	$\underline{7}$	7.5	8.5	11	14
	Y_a	8	$\underline{6}$	5	4	3	2

---

① Scitovszky T. A Reconsideration of the Theory of Tariffs[J]. Review of Economic Studies, 1941,9(11):93-94.

续表

	Commodities	Quantities				
B	X_b	2	<u>3</u>	4.5	6.5	8	12
	Y_b	5	<u>4</u>	3	2	1.5	1
Community	X	8.5	<u>10</u>	12	15	19	26
	Y	13	<u>10</u>	8	6	4.5	3

Let us look at the underlined figures. Two different distributions are illustrated with 10 units of X and Y each. In Distribution I, A as well as B get 5 units of each commodity. In Distribution II, A gets 7 units of X and 6 units of Y whereas B gets 3 of X and 4 of Y. As relative prices change,— neither do relative prices need be the same for both distributions nor do total expenditures need be constant throughout either indifference schedule—these two different distributions lead to two different community indifference schedules (results of adding the individual indifference schedules together for each distribution). These two community indifference schedules, shown in Figure 6, will yield two curves which cross each other at the point (10,10).

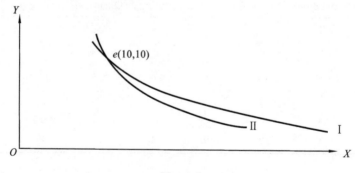

Figure 6

It may be noted that in Distribution II, A's expenditure has a greater weight in the community than B's. As it is assumed that A has a greater preference for wine and that the proportion of expenditure is in favor of wine as real income increases, more wine would be bought in Distribution II than in Distribution I at the same relative prices.

In fact, the community indifference curves may cross at any point if we care to redistribute the goods represented at that point, and through that point there will be as many community indifference curves as there are possible distributions. ①

Let us examine the significance of the intersection of the community indifference curves. In Figure 7, the significant region of our community indifference diagram has been greatly magnified. The community indifference curves for Distribution Ⅰ and Ⅱ are labelled by ab and cd respectively. e is the point of intersection of these two curves.

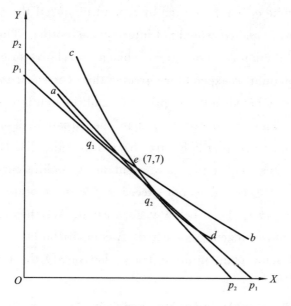

Figure 7

According to Distribution Ⅰ, any point above (or below) ab will represent a better (or worse) position than any point on ab. That is to say, the goods represented by a point above (or below) ab can be so redistributed as to make everyone better (or worse) off than the goods represented by a point on ab. Therefore, we say ce (except the point e) is

---

① For construction of the community indifference map, reference may be made to Baumol W J. The Community Indifference Map: a Construction[J]. Review of Economic Studies, 1949—1950, 17 (44): 189. Baumol's construction of the community indifference curve is more roundabout and cumbersome (and, of course, more general) since he does not directly assume a price ratio between X and Y. Instead, he works out the price ratio from the two individuals' indifference curves by equalizing one's marginal rate of substitution to the other's. His article also provides an algebraic argument for the construction of the community indifference schedules with m commodities and n persons.

superior to *ab*, and *de* is inferior to *ab*. Similarly, for Distribution II, *eb* is superior to *cd* and *ae* is inferior to *cd*.

Suppose Situation I corresponds to some point on *ab* and Distribution I is the actual distribution in Situation I (D_1). Similarly, Situation II corresponds to some point on *cd* and Distribution II is the actual distribution in Situation(D_2). Does the criterion $\sum p_2 q_2 > \sum p_2 q_1$ tells us whether q_2 is represented by a point in the segment *ce* or by a point in the segment *de* while q_1 is represented by a point on *ab*? If there is no change in distribution, *ab* and *cd* would not intersect each other. Thus, q_2 must be a point in the section *ce*. For, after a change in relative prices from p_1 to p_2, with the amount of expenditure greater than (or equal to) $\sum p_2 q_1$, the point of tangency between the p_2 price line and the indifference curve must be above *ab*. This is no different than the situation described for the individual(compare chapter 2, Figure 2b). On the other hand, if a change in distribution takes place, the new community indifference curve may (especially when $\sum p_2 q_2$ does not exceed $\sum p_2 q_1$ by too great an amount) run from above *ab* to down below it such as *cd*. We then cannot be sure where the point of tangency is located. A comparison between some point in *de*(q_2) and some point in *ab*(q_1) may also satisfy the relation $\sum p_2 q_2 > \sum p_2 q_1$.

3.12 The Double Index Number Criteria

While $\sum p_2 q_2 > \sum p_2 q_1$ does not ensure that $U(q_2) > U(q_1)$ at D_1, there is a way to ensure that $U(q_2) \not> U(q_1)$ at D_1.

Just as $\sum p_2 q_2 > \sum p_2 q_1$ means that $U(q_2) > U(q_1)$ at D_2, so $\sum p_1 q_1 > \sum p_1 q_2$ can be taken to mean that $U(q_1) > U(q_2)$ at D_1. Since we have shown that $U(q_2) > U(q_1)$ at D_2 fulfills the requirement that in the redistribution of q_1 no one can gain without imposing a loss on someone else, it can be also shown that $U(q_1) > U(q_2)$ at D_1 fulfills the requirement that in the redistribution of q_2 no one can gain without involving some other's loss. This means that if $\sum p_2 q_2 > \sum p_2 q_1$ then $U(q_1) \not> U(q_2)$ at D_2. Also, if $\sum p_1 q_1 > \sum p_1 q_2$ then $U(q_2) \not> U(q_1)$ at D_1. Then, if both $\sum p_2 q_2 > \sum p_2 q_1$ and $\sum p_1 q_1 > \sum p_1 q_2$ are obtained, the Scitovszky condition will not be satisfied.

However, it is important to note that, even if both $\sum p_2 q_2 > \sum p_2 q_1$ and $\sum p_1 q_2 > \sum p_1 q_1$ are obtained, it does not follow that $U(q_2) > U(q_1)$ at both D_1 and D_2. Just as for an individual, $\sum p_1 q_2^* > \sum p_1 q_1^*$ does not ensure that $U(q_2^*) > U(q_1^*)$, so for the community, $\sum p_1 q_2 > \sum p_1 q_1$ does not ensure that $U(q_2) > U(q_1)$ at D_1. For Lq, as shown earlier, has an upward bias in its relation to unity.

For an individual, $\sum p_2 q_2^* > \sum p_2 q_1^*$ entails $\sum p_1 q_2^* > \sum p_1 q_1^*$, but this does not hold for a community. In explaining these relations for an individual, we have introduced above the hypothetical bundle of goods $\bar{q}_2^*$ which is defined as the bundle of goods which would have been bought at p_1 prices such that $U(\bar{q}_2^*) = U(q_2^*)$. Since $U(q_2^*) > U(q_1^*)$ can be derived from the relation $\sum p_2 q_2^* > \sum p_2 q_1^*$, we have $\sum p_1 \bar{q}_2^* > \sum p_1 q_1^*$. Therefore, $\sum p_1 q_2^* > \sum p_1 \bar{q}_2^* > \sum p_1 q_1^*$. But, for the community, $\bar{q}_2$ would be required to mean a bundle of goods making everyone as well off as he is in Situation II if $\sum p_1 q_2 > \sum p_1 \bar{q}_2$ is sought. But at the same time, $\sum p_1 \bar{q}_2 > \sum p_1 q_1$ only when $\bar{q}_2$ is required to make everyone better off than he is in Situation I. We can establish either $\sum p_1 q_2 > \sum p_1 \bar{q}_2$ or $\sum p_1 \bar{q}_2 > \sum p_1 q_1$, but not both— the relation $\sum p_2 q_2 > \sum p_2 q_1$ does not show that $\bar{q}_2$ can be defined as $U(q_2) = U(\bar{q}_2)$ and $U(\bar{q}_2) > U(q_1)$ at the same time. For this reason, we need to set up Lq and Pq as double index number criteria for an increase of real national income.①

The possible case of having both $\sum p_2 q_2 > \sum p_2 q_1$ and $\sum p_1 q_1 > \sum p_1 q_2$ together may be shown in Figure 7. Thus, according to the Hicks' condition, the point q_2 is both better and worse than the point q_1 because q_2 is outside q_1 at p_2 prices and q_1 is outside q_2 at p_1 prices. This paradox is avoided by adopting the Scitovszky condition. The double index number criteria, though falling short of ensuring the fulfillment of the Scitovszky

---

① It may be helpful to note the condition, in the discussion of constant marginal proportion of expenditure in ch. 2, sec. 7, in which $Lp^* > \overset{*}{P}p$ holds for an individual. If $Lp^* > \overset{*}{P}p$ holds for every individual, $Lp > Pp$ holds for the community as a whole. From the relationship between price and quantity index numbers we know that if $Lp > Pp$, $Lq > Pq$. Then, if $Pq > 1$, $Lq > 1$.

condition, protects us from arriving at such a paradoxical conclusion. ①

3.13 Tastes, Distribution and Cost of Production

The difference in relative marginal valuations or preference, between two situations may be the result of changes in tastes, distributions, or both. The double criteria not only take care of the change in preference due to distribution but also the change in preference due to tastes. It is possible that had tastes remained constant, the mere difference in distribution would not have led to the result of having $Pq>1$ and $Lq<1$ together, though this result is actually obtained. Again, assuming distribution to be fixed, the fact that both $Pq>1$ and $Lq<1$ informs us that there is a change in tastes, for otherwise consumer's behavior is inconsistent.

Of course, the relative prices may also change as a result of changes in

---

① We may recall that the double index number criteria have long been proposed by Pigou in his Economics of Welfare. However, there is a difference. What Pigou really intended to do with the double criteria was not fully shown. He said that by assuming either that the tastes and distribution of Period I prevail in both periods or that the tastes and distribution of Period II prevail in both periods, if we have both Lq and Pq greater than unity, then we know that more than the collection of the goods of Period I can be bought in Period II while only less than the collection of the goods of Period II can be bought in Period I. Therefore, real national income is necessarily higher in Period II than in Period I. Pigou seems to forget his assumption in arriving at his conclusion. For, if more than the collection of the goods of Period I can be bought in Period II, as long as tastes and distribution are the same in both, the collection of the goods of Period II must be preferred by every member of the community to the collection of the goods of Period I. It follows that, as the collection of the goods of Period II is actually not bought in Period I, it cannot be bought in Period I. Similarly, if only less than the collection of the goods of Period II can be bought in Period I, it follows necessarily that more than the collection of the goods of Period I can be bought in Period II. For this reason, Ragnar Frisch wanted the Pigou's criteria to be sharpened into: if $Pq>1$, real national income increases ($Lq>1$ is superfluous because it is already implied by $Pq>1$); if $Lq<1$, real national income decreases ($Pq<1$ is superfluous because it is already implied by $Lq<1$). See R. Frisch, op. cit., pp. 20-21.

In using his double criteria, perhaps, Pigou meant to say that if tastes and distribution in Period I are the same as prevailed in Period II, $Pq>1$ would mean that we can buy in Period II more than the collection of the goods of Period I, therefore real national income is higher in Period II than in Period I; if tastes and distribution in Period I are different from those prevailing in Period II, $Lq>1$ would mean that we can buy in Period II more than the collection of the goods of Period I. Only then, he would come close to our analysis. However, though $Pq>1$ means $U(q_2)>U(q_1)$ for the tastes and the distribution of Period II, $Lq>1$ does not ensure $U(q_2)>U(q_1)$ for the tastes and the distribution of Period I. In addition, it is not satisfactory to say that we can buy in Period II more than the collection of the goods in Period I in either case.

the cost of production even if preference proves unchanged. But, in this case, if $Pq>1$, then $Lq>1$; if $Lq<1$, then $Pq<1$. For, though $Pq>1$ does not mean that we can produce q_2 as well as q_1 in Situation Ⅱ, it does indicate $U(q_2)>U(q_1)$. As $U(q_2)>U(q_1)$, $Pq>1$ implies $Lq>1$. By analogy, $Lq<1$ implies both $U(q_1)>U(q_2)$ and $Pq<1$.

3.14 The Utility Possibility Function

The fulfillment of the Scitovszky condition assures us that one collection of goods is better than the other with either of the two actual distributions. We still do not know, however, whether one collection of goods is better than the other according to a third distribution or according to any alternative possible distribution. Therefore, to accept the Scitovszky condition of increase of real national income would imply a preference for the actual distributions against other possible or potential distributions. ①

In view of this defect, Samuelson defines the utility possibility function. This function states that the difference between the utilities of two bundles of goods(from the point of view of making everyone better off or worse off) is a function of a distribution which may assume an infinite number of forms. Only when the difference is always positive or negative for all forms of distribution, can we say one bundle of goods is better or worse than the other. In other words, q_2 is said to be better than q_1 only when, no matter how we distribute q_1, there is always a corresponding distribution of q_2 which makes everybody better off in Situation Ⅱ than he was in Situation Ⅰ. ②

How can we incorporate this utility possibility function into our index number criteria?

We have shown that through any given point there can be drawn as many community indifference curves as there are distributions. As each

---

① To this, however, Scitovszky already answered: "We are not interested in all possible welfare distributions. There are only two distributions of welfare that really matter. Those actually obtaining immediately before and after the change contemplated." See his "A Note on Welfare Propositions in Economics," p.86.

② Samuelson P A. Evaluation of Real National Income[J]. Oxford Economic Papers, 1950, 2(1): 5.

community indifference curve has a different slope at this point and this slope corresponds to the price line, we say that for each distribution there is a different price line. This means that any point will be valued differently according to the distribution used. According to the utility possibility function, the number of index number criteria for an increase of real national income is therefore infinite. Let there be m possible sets of relative prices, each corresponding to a different distribution. The index number criteria would (if q_2 is shown to be better than q_1) run as follows:

$$\sum p_1 q_2 > \sum p_1 q_1,$$
$$\sum p_2 q_2 > \sum p_2 q_1,$$
$$\sum p_3 q_2 > \sum p_3 q_1,$$
$$\cdots\cdots$$
$$\sum p_m q_2 > \sum p_m q_1.$$

In diagrammatical terms, the point q_2 is outside the point q_1 for all possible price lines. This condition cannot be satisfied unless there is more of every good and less of none in q_2 than in q_1. Thus, the consideration of the utility possibility function makes real national income comparison extremely restrictive.

However, the application of the double criteria is not so simple as it appears to be as more and more situations are considered. According to the utility possibility function, we can say that one situation is better than another only when there is more of every good in one situation than in another. Then, if Situation Ⅰ is better than Situation Ⅱ and Situation Ⅱ is better than Situation Ⅲ, Situation Ⅰ is necessarily better than Situation Ⅲ because there will be more of every good and less of none in Situation Ⅰ than in Situation Ⅲ. But the double criteria will not enable us to make such a transitive comparison.

Consider the three points q_1, q_2 and q_3 in Figure 8, their respective price lines being denoted by p_1, p_2 and p_3. That is, q_1, q_2 and q_3 are actually bought respectively at p_1, p_2 and p_3 prices. By drawing through each point the other two's price lines, we see that

$$\sum p_1 q_1 > \sum p_1 q_2 \text{ and } \sum p_2 q_1 > \sum p_2 q_2,$$
$$\sum p_2 q_2 > \sum p_2 q_3 \text{ and } \sum p_3 q_2 > \sum p_3 q_3,$$

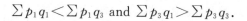

$\sum p_1 q_1 < \sum p_1 q_3$ and $\sum p_3 q_1 > \sum p_3 q_3$.

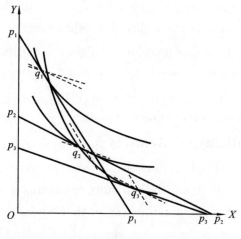

Figure 8

According to the first two double criteria, we may say that q_1 is better than q_2 (from the chart we know also that the Scitovszky condition is satisfied because q_1 is above the indifference curve of q_2, and q_2 is below the indifference curve of q_1) and q_2 is better than q_3 (the Scitovszky condition is again satisfied). But, shall we conclude that q_1 is therefore better than q_3? From the last double criteria we know that q_1 is indifferent to q_3, since each point is worse than the other at the other's distribution (the Scitovszky condition is not satisfied). ①

Thus, in comparing m points, even if we know according to our double criteria that the point q_i is better than all the other points and q_{m+1} is better than q_i,② we cannot infer than q_{m+1} is therefore better than all the other m

---

① A slightly different numerical example to show the intransitiveness of the Scitovszky condition can be found in Arrow, op. cit., pp. 44-45.

② That is,

$$\sum p_i q_i > \sum p_i q_1, \sum p_1 q_i > \sum p_1 q_1, \text{and} \sum p_i p_i < \sum p_i q_{m+1},$$
$$\sum p_i q_i > \sum p_i q_2, \sum p_2 q_i < \sum p_2 q_2, \sum p_{m+1} q_i < \sum p_{m+1} q_{m+1}.$$
$$\cdots\cdots$$
$$\sum p_i q_i > \sum p_i q_{i-1}, \sum p_{i-1} q_i > \sum p_{i-1} q_{i-1},$$
$$\sum p_i q_i > \sum p_i q_{i+1}, \sum p_{i+1} q_i > \sum p_{i+1} q_{i+1},$$
$$\cdots\cdots$$
$$\sum p_i q_i > \sum p_i q_m, \sum p_m q_i > \sum p_m q_m.$$

points. We must apply the double criteria between q_{m+1} and each of the other m points. Thus, the processes come closer to the ones described under the consideration of the utility possibility function.

It is true that the utility possibility function has little claim to practical application. It merits our consideration because it strongly reminds us of the inadequacy and limitations of the principle of compensation as applied to the measurement of real national income.

3.15 Potentiality vs. Actuality

The principle of compensation, as first conceived by Kaldor and Hicks, was intended to avoid the difficulty of making a value judgment of distribution through using potential rather than actual real national income comparisons. But, as we have seen above, the Kaldor-Hicks version[①] and even the Scitovszky amendment to the definition of increase of real national income, strictly speaking, does involve an unintended value judgment in the sense of preferring some distributions to others. To avoid giving any preference whatsoever, we have to make real national income comparison by considering the utility possibility function. Then, only the fulfillment of the condition that there are more of all goods in one situation than in another enables us to say that one situation is better than the other. Consequently, the condition of increase of potential real national income would deprive the principle of compensation of its practical import for economic policy. For the condition that there is more of every good in one situation than in another may never be found in reality.

Even if everyone is actually better off in one situation than in another, we cannot say, according to the utility possibility function, that the potential real national income is greater in the former than in the latter. Truly, it is doubtful if we can say that the actual real national income is greater, say, in Situation Ⅱ than in Situation Ⅰ if <u>anyone</u> is worse off in Situation Ⅱ than in Situation Ⅰ, because a dubious interpersonal comparison of utilities is then involved. But, assuming that $\sum p_2 q_2^* > \sum p_2 q_1^*$

---

① Note Little's remark that "What Mr. Kaldor did, in fact, was to propose a definition of 'increase of wealth' which ignored distribution." See his A Critique of Welfare Economics, p. 93.

holds for everyone, we are assured that everyone is better off in Situation Ⅱ than in Situation Ⅰ. Hence, the actual real national income has increased from Situation Ⅰ to Situation Ⅱ. However, the fact that $\sum p_2 q_2^* > \sum p_2 q_1^*$ holds for everyone does not ensure that there are more of all goods and less of none in Situation Ⅱ than in Situation Ⅰ. On the other hand, it is obvious that more of all goods for the community do not guarantee everyone to be better off if there is a change in distribution. ①

4. The Valuation of Real Income for a Group of Individuals (continued)

Welfare economists often regard potential economic welfare as synonymous with productive efficiency (or productivity) in justifying their attempt to measure the change of real national income irrespective of the actual change in distribution. ② In this chapter, the possible divergence between the magnitude "economic welfare" and the magnitude "productive fficiency" due to imperfections of the market will not be considered. ③ Rather, considerations will be given to the possible use of aggregate quantity index numbers for comparing productive efficiencies between different situations. Although interpretation of changes in real national income in terms of productive efficiency may be desirable because it eliminates the necessity for interpersonal comparisons of utility, from the

---

① Since both the conditions of increase of potential real national income and of increase of actual real national income, which are completely freed from value judgments, can hardly be found in reality, there is a strong suggestion that an appraisal of distribution is necessary for real national income comparison.

② For example, see Hicks, "The Rehabilitation of Consumer's Surplus," p. 111; Scitovszky T. Welfare and Competition [M]. Chicago: Irwin, 1951: 71; Pigou A G. Income, An Introduction to Economics[M]. London: MacMillan, 1948: 15-16.

③ See, for example, Hicks, "The Valuation of the Social Income," p. 122, "If competition were perfect, and if state activities were so designed as not to disturb the optimum organization of production, marginal utilities and prices and marginal costs would all be proportional, so that the same valuation which gave us the social income as a measure of economic welfare would also give us the social income as a measure of productivity. It is the departure of the system from the optimum, whether as a result of indirect taxation or as a result of imperfect competition, which upsets the equivalence and makes the measurement of economic welfare a different thing from the measurement of productivity."

The problem of imperfect economy will be taken up in the following chapter.

standpoint of the present investigation the important thing is whether it is possible to tell, with a mere knowledge of index numbers, whether there has been an increase(or decrease)of productive efficiency.

4.1 Revenue Maximization and Index Number Biases

Whereas, at given prices, a consumer tries to minimize his expenditure for a given amount of utility, a producer tries to maximize his revenue at a given cost. For the economy as a whole, though the total expenditure is always identical with the total revenue, the minimization effort on the part of the consumers and the maximization effort on the part of the producers, while not producing any excess of revenue over expenditure, will bring about a national income of maximum utility with minimum cost.

The maximization problem of the productive economy may be compared to that confronted by a firm producing joint products. Let us assume that there were no business taxes and that conditions of perfect competition prevailed. For a firm producing joint products, the amounts of the various products will be so adjusted that their marginal costs are proportional to their prices. Assuming increasing marginal rate of substitution for products(i. e. increasing costs)over the relevant range of production,[1] the iso-cost curve (showing what combinations of the various products would be produced at a given cost)will be concave to the origin in the neighborhood of the point of equilibrium. By varying the cost of production under constant productive efficiency, iso-cost contours can be drawn. For the productive economy as a whole, the totality of the national resources may be treated as a given cost and various combinations of national output which would be produced in the economy form the iso-cost curve. But unlike the productive resources(input)of a firm, the totality of the national resources can hardly be treated as a variable. If we want to construct an iso-cost contour map for the economy as a whole, we would have a one-curve map. However, we may treat the iso-cost curve as an iso-

---

[1] See Hicks J R. Value and Capital[M]. 2nd ed. Oxford: Clarendon, 1946: 87, footnote 1, "Increasing marginal rate of substitution for products, because the total value of products secured has to be maximized. "If this rate is diminishing(or constant), the condition would be unstable.

efficiency curve at a given cost, i. e. at the cost of the totality of the national resources. Then, by changing the state of productive efficiency, each time corresponding to a given situation, we have an iso-efficiency contour map in which each curve represents a different state of productive efficiency.

Let us consider only one of the iso-efficiency curves, which is supposed to describe the state of productive efficiency in Situation I. In Figure 9, X and Y refer to two commodities and the iso-efficiency curve is denoted by EE'. If the price line is bb', q_1 will be produced; if the price line is dd', q_2 will be produced. Through q_1 we draw cc' parallel to dd' and through q_2, aa' parallel to bb'. Then, we have $od/oc > 1$ and $oa/ob < 1$. As oa, ob, oc and od represent the expenditures $\sum p_1 q_2$, $\sum p_1 q_1$, $\sum p_2 q_1$ and $\sum p_2 q_2$ respectively, we have, by substitution, $\sum p_2 q_2 / \sum p_2 q_1 > 1$ and $\sum p_1 q_2 / \sum p_1 q_1 < 1$.

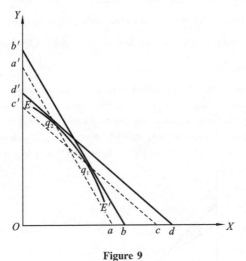

Figure 9

Let productive efficiency be a function of quantities of production: $E = E(q)$, we have $E(q_2) = E(q_1)$, or $E(q_2)/E(q_1) = 1$ as both q_2 and q_1 correspond to the same level of productive efficiency. Therefore, when used as an index of productive efficiency, Lq has a downward bias whereas Pq has an upward bias for the same productive efficiency.

Due to the concavity of the iso-efficiency curve, if $Lq > 1$, q_2 must lie above the iso-efficiency curve of which q_1 is a point. If $Pq<1$, q_1 must lie above the iso-efficiency curve of which q_2 is a point. ①

Under the assumption that the iso-efficiency curves do not intersect, if q_2 lies above the iso-efficiency curve of q_1, q_1 must lie below the iso-efficiency curve of q_2 and, also, every point on the iso-efficiency curve of q_1 will lie below the iso-efficiency curve of q_2. Conversely, if q_1 lies above the iso-efficiency curve of q_2, q_2 must lie below the iso-efficiency curve of q_1 and, also every point on the iso-efficiency curve of q_2 will lie below the iso-efficiency curve of q_1. Then, we are able to say that, if $Lq>1$, $E(q_2)>E(q_1)$; if $Pq<1$, $E(q_1)>E(q_2)$. Furthermore, it can be shown that $Lq>1$ entails $Pq>1$, and $Pq<1$ entails $Lq<1$. ②

4.2 Restrictions on Interpretation of Results

However, changes in technology and productive conditions often impede the production of some commodities while facilitating the production of others. As a result, the substitution function for products may change greatly from one situation to another. Of the two points under

---

① Since it is less likely that the iso-efficiency curve is uniformly concave to the origin than that the utility indifference curve is uniformly convex to the origin, we must add that the index number criteria for the relative positions of productive efficiency hold only when one point is in the neighborhood of the other. It can be shown that, as the two points fall apart, each point is above the other point's price line. The theory that Lq has a downward and Pq has an upward bias no longer applies. See Figure 10.

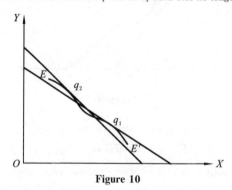

Figure 10

② The processes by which we have shown that $Pq>1$ entails $Lq>1$ in the utility index apply in analogous ways here. Since $Lq>1$ means $R(q_2)>R(q_1)$, by introducing $\bar{q}_1$ as a combination of goods which would have been produced at p_2 prices on the iso-efficiency curve of q_1, we have $\sum p_2 q_2 > \sum p_2 \bar{q}_1 > \sum p_2 q_1$, i.e. $Pq>1$.

comparison, unless one lies far above the other, it is not unlikely that the iso-efficiency curves of these two points intersect each other. ①

As the iso-efficiency curves may intersect, we are deprived of the right of rating one point above another in terms of productive efficiency by any such simple index number criteria as discussed above. The reason for this can be demonstrated in two steps: first, $Lq>1$ tells us only that q_2 lies above the iso-efficiency curve of q_1, but not that q_1 lies below the iso-efficiency curve of q_2; whereas $Pq<1$ tells us only that q_1 lies above the iso-efficiency curve of q_2, but not that q_2 lies below the iso-efficiency curve of q_1. Second, even if we know, e. g., that q_2 lies above the iso-efficiency curve of q_1 and also q_1 lies below the iso-efficiency curve of q_2, (the case in which there is more of every good in q_2 than in q_1) we still do not know whether or not every point on the iso-efficiency curve of q_1 lies below the iso-efficiency curve of q_2, in other words, we still do not know that the iso-efficiency curve of q_2 lies uniformly outside that of q_1. Thus, our conclusions are even more restricted if we interpret real national income in terms of productive efficiency or productivity than in terms of potential utility.

However, if we are of the opinion that only the actual points are pertinent for comparative purposes, then the double criteria may again serve to tell which one point is better than the other without apparent contradiction. ② The only difference between the utility and the productive efficiency index number criteria is that while $Pq>1$ ensures that q_1 is

---

① See Kuznets S. On the Valuation of Social Income—Reflections on Prof. Hicks' Article, Part Ⅱ [J]. Economica, 1948, 15(57): 118. He furnishes an answer to the question: why we are reluctant to make the assumption that the iso-efficiency curves do not cut? For, "If we interpret such an assumption most rigidly," the iso-efficiency curves of q_1 and q_2 "would be constructed under completely unchanged conditions of technique and structure of production." Then, it would be nonsense to compare the productive efficiencies represented respectively by q_1 and q_2.

For further discussions, see Simpson P B. Transformation Functions in the Theory of Production Indexes[J]. Journal of the American Statistical Association, Vol. 46, no. 254, June 1951, 46(254): 230-231.

② Pigou A G. Income, an introduction to economics[M]. London: MacMillan, 1948: 15-16. The double criteria are again stated for testing an increase of "the income producing power of a country's resources."

below the utility indifference curve of q_2, $Lq>1$ ensures that q_2 is above the iso-efficiency curve of q_1. On the other hand, $Lq<1$ ensures that q_2 is below the utility indifference curve of q_1 and $Pq<1$ ensures that q_1 is above the iso-efficiency curve of q_2.

The concrete meaning of the index number criteria for productive efficiency is that, if $Lq>1$, we know that the same amounts of goods produced in Situation II cannot be produced in Situation I, but we are not sure whether or not the same amounts of goods produced in Situation I can be produced in Situation II. Conversely, if $Pq<1$, we know that the same amounts of goods produced in Situation I cannot be produced in Situation II, but we are not sure whether or not the same amounts of goods produced in Situation II can be produced in Situation I.

For a simple numerical example, let us consider only two commodities X and Y of which the quantities and the prices in Situation I and Situation II are as follows:

	Quantities		Prices($)		Total Revenues
	X	Y	P_x	P_y	
Situation I	4	5	1	1	9
Situation II	5	3	1.25	1.75	11.50

From these hypothetical quantity and price data, we obtain $Pq = 11.50/13.75<1$ and $Lq=8/9<1$. Since \$11.50 is the maximum revenue in Situation II, it is impossible to produce in Situation II 4 units of X and 5 units of Y which would have yielded a revenue of \$13.75 at the prices of Situation II. Whether it is possible or not, while $Lq<1$, to produce 5 units of X and 3 units of Y in Situation I depends on the marginal rate of substitution of X for Y. It is possible if this rate is 2 or less.

In terms of productive efficiency, we may say that if $Lq>1$, productive efficiency is higher in Situation II than in Situation I only insofar as the production of the actual combination of the goods of Situation II is attempted. If $Pq<1$, the converse is true.①

---

① To interpret on individual firm's basis, if $Lq>1$, the goods of Situation II can be so redistributed as to put everyone on a higher level of productivity than he was in Situation I; and conversely.

4.3 Potential Welfare in Relation to Iso-efficiency Curve

Quantities and prices are interdependent. Any change in distribution will bring about a change in community indifference schedules and therefore a change in relative prices. As a result, different combination of goods will be produced at the existing level of productive efficiency.

In Figure. 11, let EE' be the iso-efficiency curve. As a result of a redistribution of the goods represented by the point q, the community indifference curve changes from I to I' and the price line changes from p to p'. Through q we draw a lower community indifference curve $\bar{I}$ with respect to I'. After the redistribution consumers may spend less money by purchasing a bundle of goods, $\bar{q}$, different from q without thereby reducing the amount of utility. At the same time producers may acquire more revenue by producing a bundle of goods, q', different from q without increasing productive efficiency (at p' prices, $\bar{q}$ is below q and q' is above q). Therefore, to weight q by a set of relative prices different from the original prices, on the one hand exaggerates the amount of utility and on the other understates the level of productive efficiency. Thus, while it is true that $Pq > 1$ signifies that there is some redistribution of q_1 (represented by q in Figure 11) which would make everyone worse off in Situation I than he actually is in Situation II, it is not necessarily true that $Pq > 1$ signifies a possible redistribution of the goods of Situation I (represented by q' in Figure 11) <u>that</u> <u>would</u> <u>be</u> <u>produced</u> <u>if</u> <u>that</u> <u>redistribution</u> <u>were</u> <u>accomplished</u> making everyone worse off in Situation I than he actually is in Situation II.

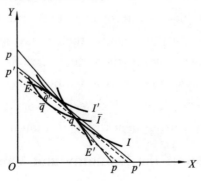

Figure 11

4.4 Limitations of Real Income Comparison for a Group of Individuals

Since aggregate quantity index numbers tell us only the total quantity of goods acquired by a group of individuals and nothing about its distribution, and since real income comparisons under conditions of changing distribution involve dubious interpersonal comparisons, the index number analysis in the last two chapters has been largely devoted to potential real income comparison. Thus, the statement $U(q_2) > U(q_1)$ means that everyone is made better off in Situation II than in Situation I. For this reason, the relation $U(q_2) > U(q_1)$ is always qualified by some given distribution. Real national income comparison in potential terms, therefore, is confined to points and cannot be extended to the whole indifference curves on which these points lie without making additional assumptions.

For the individual consumer, $Pq > 1$ entails $Lq > 1$, and $Lq < 1$ entails $Pq < 1$. For a group of consumers, these index number relationships may not hold true because of changes in income distribution. As a result, double criteria are necessary as a protection against apparently contradictory conclusions. However, since information is lacking as to how people of one situation would price the goods of another situation, it is not possible to prove that the goods of one situation are better than those of another at the other's distribution, e.g., $U(q_2) > U(q_1)$ at D_1. The most that can be inferred is that the greater the amount by which $\sum p_1 q_2$ exceeds $\sum p_1 q_1$, the more likely that $U(q_2) > U(q_1)$ at D_1.

The consideration of the utility possibility function and the interpretation of real national income in terms of productive efficiency further show the necessity of qualification in saying $U(q_2) > U(q_1)$ or $E(q_2) > E(q_1)$. In theory, when these qualifications are made, the use of index numbers as indicators of real national income attains objective validity. From a practical point of view, however, the necessity for these qualifications may leave little room for application in actual work.

5. Imperfection of the Economy and Valuation Problems

We have thus far limited our problem to the consumption goods purchased in a free market. In this chapter, we shall extend our inquiry into the valuation of real income to all goods and services which make up the net national product and for which there may or may not be a free market. Since, as will be shown, the goods of various sectors of the economy are valued on different bases or by different standards, and, also, the goods under government control or monopolistic competition are priced according to different principles, elements of incomparability are introduced into the aggregate index numbers. As a result, no uniformly simple and clear interpretation of changes in real national income is possible. In what follows, means will be sought to value all goods and services consistently and uniformly so that the resultant index numbers may be unambiguously interpreted.

5.1 Disequilibrium Prices or Valuations

The condition that marginal utilities are proportional to prices has been essential in the comparison of real income(utility). Unfortunately, in the real world, this condition is not strictly fulfilled because of the following considerations:

(1) The actual market may not be free. At market prices the consumers may not be able to buy the amounts of commodities they wish. As in the case of rationing or price control, there may be an absolute shortage of supply.

(2) There are no sale prices for government services. Taxes for the individuals may have no relation to value of the services(if any) which are received.

(3) There are capital goods which are bought for further production or for resale, not for consumption. The prices at which these goods are bought may represent their marginal productivities rather than marginal utilities to the buyers. ①

---

①　If the buyers are monopsonists, marginal productivities would be proportional to marginal outlays rather than prices. See Boulding K E. Economic Analysis (Revised ed). New York: Harper, 1948:715, "The marginal outlay is defined as the increase in total outlay which results from a unit increase in the purchases of an input."

(4) The monopolistic practice of raising price by restricting output also affect the utility index numbers in an artificial and indirect way. ①

(5) From a practical point of view, the availability and reliability of the national income data on price and quantity has a bearing on the precision of the aggregate index numbers. For example, the national income data are confined to the market activities. The subtraction of depreciation, etc., from gross national product to arrive at net national product involves judgment of a quite arbitrary nature. Differences in quality may not be adequately reduced to quantity in classification. For the same commodities there are many quoted prices, etc.

For example, when rationing is put into effect, demand for the rationed goods is artificially suppressed. At market prices the amount demanded exceeds the available supply. Hence, the marginal utilities of rationed goods, as compared with those of unrationed goods, tend to be more than proportional to prices.

Again, as there are no sale prices for government services, a comparison of marginal utilities of the various kinds of government services with those of the market goods cannot be made. The valuation of government services at cost does not help because, in the absence of a price system, an objective standard is lacking both for determining how much and where the government should spent its resources. Since the valuation of government services at cost(i. e. at factor market prices), is the only feasible procedure,② it is doubtful whether marginal utilities are proportional to prices.

It is less obvious that the existence of capital goods disturbs the proportionality between marginal utilities and prices. This possibility is not difficult to see if expenses paid for capital goods are regarded as the costs of production of future consumption goods. The marginal utilities of the capital goods are then measured not by the prices of the capital goods

---

① This point will be discussed in section 10 of this chapter; for the time being, we just note it.

② The practice, as favored by Kuznets, of valuing total governmental services on the basis of tax payments is still less realistic. See Kuznets S. National Income and Its Composition 1919—1938, Vol. 1. New York: National Bureau of Economic Research, 1941: 33.

but the prices of the services which the capital goods eventually produce for consumers (e. g. , inventory can neither be weighted by original costs nor by market prices if the real economic importance of inventory is to be determined). These latter prices are not known until sometime in the future when the capital goods are completely transformed into consumers' goods.

If the economy operated so that prices were always equal to costs, the prices of capital goods would always be the same as the prices of the services which the capital goods ultimately produce—after allowing for interest charges. Such perfection, however, is not attained except perhaps in a stationary state. In a dynamic world things are changing. Due to the fact that most productive factors cannot be transferred from one use to another without loss, the prices of capital goods are governed by the expected prices of their future services. If the expected prices are wrong, the present prices would not correctly represent the marginal utilities of the services which the capital goods will produce. ① Due to imperfect knowledge of the future, expectations are usually not fulfilled. Equilibrium overtime, to use Hicks' term, is not achieved. ② As a rule, some capital goods are valued too high and others too low in comparison to the valuation of consumption goods. ③

Finally, if the price and quantity data are not correct, proportionality between marginal utilities and prices would be an accident.

5.2 Qualifications of the Index Number Criteria

If disequilibrium prices, or valuations, which are not proportional to marginal utilities, enter into our quantity index numbers as weights, the criteria which we have developed for increase of real income or utility cannot hold without reservations. These criteria will probably not be affected if the disequilibrium weights exaggerate or understate the amount

---

① See Appendix A to this chapter for a supporting argument.
② Hicks J R. Value and Capital[M]. 2nd ed. Oxford: Clarendon, 1946: 132.
③ It may be added that the purchase of the consumers' durable goods also involves an expectation since there may be a considerable time interval between the act of purchasing and the act of consuming.

of real income to the same extent in both situations, that is, if it is found in both situations that the role and the efficiency of the government are much the same; that the fulfillment of the producers' expectation is much the same; that the shortage of supply, if any, at the market prices is much the same, or that any difference in one respect is just compensated by a reverse difference in the other. However, such a coincidence of sameness or of difference cannot be rigorously demonstrated. Moreover, if it is apparent that one situation is much different from another (for example, if one situation is under general rationing and the other is not),[①] we can no longer rely on our index number criteria without applying a correction.

5.3 The "Virtual Price System"

If the prices or valuations of the goods used as weights do not represent their marginal utilities, the price line, in terms of our indifference analysis of two goods, will intersect the indifference curve. $\sum p_2 q_2 > \sum p_2 q_1$ will not ensure that the point q_1 falls below the indifference curve of which q_2 is a point. Thus, it is not necessarily true that, for given tastes and distribution, $U(q_2) > U(q_1)$.[②] In order to show $U(q_2) > U(q_1)$, we need to find a set of relative prices at which q_2 would represent an optimum—that is, where the price of each good correctly expresses its marginal utility (at which prices, demand would equal supply for the rationed goods; the government would find just enough buyers for its services if the latter are put on the market, and, expectation would coincide with realization for the capital goods). This set of equilibrium prices may be called the "virtual price system" after Rothbarth.[③]

Let $\bar{p}_2$ denote the virtual prices in Situation II. If $\sum \bar{p}_2 q_2 > \sum \bar{p}_2 q_1$, then $U(q_2) > U(q_1)$. It is our task now to find the value of $\sum \bar{p}_2 q_2 / \sum \bar{p}_2 q_1$ or, for short, $\bar{P}q$.

---

① For an example of the actual problems to be considered in the valuation of real national income in a controlled economy, see Clark Colin. The Conditions of Economic Progress[M]. 2nd ed. London: MacMillan, 1951: 163.

② Hence, $\sum p_2 q_2 > \sum p_2 q_1$ does not entail $\sum p_1 q_2 > \sum p_1 q_1$.

③ Rothbarth E. The measurement of changes in real income under conditions of rationing[J]. Review of Economic Studies, 1940-1941, 8.

$\overline{Pq}$ cannot be directly calculated until the virtual prices, $\overline{p}_2$, are found. However, it would be quite difficult to obtain an exact estimate of $\overline{p}_2$. The difficulty of estimating $\overline{p}_2$ for government services and for capital goods, in the sense we require, is much greater than that for rationed goods. For this reason, it is desirable to adopt a simple device so that under fortunate circumstances, a mere knowledge of the direction in which $\overline{p}_2$ differ from p_2 will enable us to determine $\overline{Pq}$ in terms of "greater than" or "less than" one.

5.4 Price-quantity Relationship

If tastes and distribution remain unchanged, we would expect an inverse relationship between quantity and price. That is, prices tend to rise most or fall least for those goods whose quantities fall most or rise least, and vice versa. Thus, the Paasche's quantity index number

$$Pq = \frac{\sum p_2 q_2}{\sum p_2 q_1} = \frac{\sum p_2 q_2}{\sum \frac{q_1}{q_2} p_2 q_2} = \frac{1}{\dfrac{\sum \frac{q_1}{q_2} p_2 q_2}{\sum p_2 q_2}},$$

yields a lower value the more inverse the relationship between quantity and price because a greater weight will be given to the high quantity relative q_1/q_2 and a lesser weight will be given to the low quantity relative q_1/q_2.

To illustrate this proposition, let us consider two commodities X and Y of which q_1, q_2 and p_2 assume the following numerical values according to three different degrees of inverseness.

	Commodities	q_1	q_2	q_1/q_2	p_2	$p_2 q_2$	$\frac{q_1}{q_2} p_2 q_2$	
(1)	X	20	10	2	5	50	100	$Pq=1$
	Y	10	20	1/2	5	$\frac{100}{150}$	$\frac{50}{150}$	
(2)	X	20	10	2	4	40	80	$Pq>1$
	Y	10	20	1/2	6	$\frac{120}{160}$	$\frac{60}{140}$	

	Commodities	q_1	q_2	q_1/q_2	p_2	p_2q_2	$\frac{q_1}{q_2}p_2q_2$	
(3)	X	20	10	2	20	200	400	$Pq<1$
	Y	10	20	1/2	10	$\overline{\frac{200}{400}}$	$\overline{\frac{100}{500}}$	

p_2 and q_2 are less inversely related in (2) than in (1), and in (1) than in (3), and, Pq is greatest in (2) and least in (3). That is, Pq is less the more inversely related are p_2 and q_2.

On the other hand, the Laspeyres' quantity index number,

$$Lq = \frac{\sum p_1 q_2}{\sum p_1 q_1} = \frac{\sum \frac{q_2}{q_1} p_1 q_1}{\sum p_1 q_1},$$

yields a higher value the more inverse the relationship between quantity and price because a greater weight will be given to the high relative q_2/q_1 and a lesser weight will be given to the low relative, q_2/q_1.

Let $\overline{p}_2$ and $\overline{p}_1$ denote the virtual price systems for q_2 and q_1 respectively. Thus, $\overline{Pq}$ is greater or less than Pq according as $\overline{p}_2$ and q_2 are less or more inversely related than p_2 and q_2, and $\overline{Lq}$ is greater or less than Lq according as $\overline{p}_1$ and q_1 are more or less inversely related than p_1 and q_1.

5.5 The Direction in Which Virtual Prices Differ from Actual Prices

But how can we tell the degree of the inverse relationship between quantity and price generally when more than two commodities are to be considered? We have said above that the greater the inverseness of the price-quantity relationship, in the greater the weights given to the high quantity relatives (q_1/q_2 in the Paasche's and q_2/q_1 in the Laspeyres') and the lesser the weights given to the low quantity relatives. But how high a relative should be regarded as high and how low as low? What is the dividing line?

For the present purpose, we define q_1/q_2 as a high or low relative according to where

$$\frac{q_1}{q_2} > \frac{\sum p_2 q_1}{\sum p_2 q_2} \quad \text{or} \quad \frac{q_1}{q_2} < \frac{\sum p_2 q_1}{\sum p_2 q_2},$$

and, analogously, q_2/q_1 is high or low according to where

$$\frac{q_2}{q_1} > \frac{\sum p_1 q_2}{\sum p_1 q_1} \quad \text{or} \quad \frac{q_2}{q_1} < \frac{\sum p_1 q_2}{\sum p_1 q_1}.$$

Let us consider q_1/q_2 first, and write $\overline{p_2} - p_2 = p'_2$. If $p'_2 < 0$ for those high quantity relatives and $p'_2 > 0$ for those low quantity relatives, we have, for the high relatives (symbolized with subscript h),

$$\frac{\sum p'_2 q_2}{\sum \left(\frac{q_1}{q_2}\right)_h p'_2 q_2} < \frac{\sum p_2 q_2}{\sum \frac{q_1}{q_2} p_2 q_2},$$

hence, remembering that $p'_2 < 0$ and that q_1, q_2 and p_2 are non-negative,

$$\frac{\sum p_2 q_2 + \sum p'_2 q_2}{\sum \frac{q_1}{q_2} p_2 q_2 + \sum \left(\frac{q_1}{q_2}\right)_h p'_2 q_2} > \frac{\sum p_2 q_2}{\sum \frac{q_1}{q_2} p_2 q_2},$$

if $\sum p_2 q_1 > |\sum (q_1/q_2)_h p'_2 q_2|$. ① For the low relatives (symbolized with subscript l),

$$\frac{\sum p'_2 q_2}{\sum \left(\frac{q_1}{q_2}\right)_l p'_2 q_2} > \frac{\sum p_2 q_2}{\sum \frac{q_1}{q_2} p_2 q_2},$$

hence,

$$\frac{\sum p_2 q_2 + \sum p'_2 q_2}{\sum \frac{q_1}{q_2} p_2 q_2 + \sum \left(\frac{q_1}{q_2}\right)_l p'_2 q_2} > \frac{\sum p_2 q_2}{\sum \frac{q_1}{q_2} p_2 q_2}.$$

Combining both the high and the low relatives, we have, putting $p''_2 = p'_2$ where $p'_2 < 0$,

$$\frac{\sum p_2 q_2 + \sum p''_2 q_2 + \sum p'_2 q_2}{\sum \frac{q_1}{q_2} p_2 q_2 + \sum \left(\frac{q_1}{q_2}\right)_h p''_2 q_2 + \sum \left(\frac{q_1}{q_2}\right)_l p'_2 q_2} > \frac{\sum p_2 q_2}{\sum \frac{q_1}{q_2} p_2 q_2},$$

that is, $\overline{P}q > Pq$.

By analogous processes, it can be shown that if $p'_2 < 0$ for those low quantity relatives and $p'_2 > 0$ for those high quantity relatives, we have $\overline{P}q < Pq$.

---

① To work in a more simplified form, let a, b, c and d be positive, real numbers and $a' = -a$ and $b' = -b$. If $a/b < c/d, -a/-b < c/d$, i.e. $a'/b' < c/d$. Since b' is negative, multiplying both sides of the inequality by b', we have $a' > b'c/d$. Then, adding c to both sides, we have $a' + c > b'c/d + c$, that is, $a' + c > (b' + d)c/d$. Again, multiplying both sides by $1/(b' + d)$, we have $(a' + c)/(b' + d) > c/d$ if $b' + d > 0$, or $d > |b'|$; and, $(a' + c)/(b' + d) < c/d$ if $b' + d < 0$, or $d < |b'|$.

Let $\bar{p}_1 - p_1 = p'_1$. Again, it can be shown that if $p'_1 < 0$ for those low q_2/q_1 and $p'_1 > 0$ for those high q_2/q_1, we have $\overline{L}q > Lq$; if $p'_1 < 0$ for those high q_2/q_1 and $p'_1 > 0$ for those low q_2/q_1, we have $\overline{L}q > Lq$.

Thus, if $Pq > 1$ and $\overline{P}q > Pq$, $\overline{P}q > 1$ (if $Lq < 1$ and $\overline{L}q < Lq$, $\overline{L}q < 1$); if $Pq < 1$ and $\overline{P}q < Pq$, $\overline{P}q < 1$; if $Lq > 1$ and $\overline{L}q > Lq$, $\overline{L}q > 1$. However, if $Pq < 1$ and $\overline{P}q > Pq$, we have $\overline{P}q \lesseqgtr 1$ (if $Lq > 1$ and $\overline{L}q < Lq$, we have $\overline{L}q \lesseqgtr 1$).

5.6 Possible Applications

To determine whether $\overline{P}q > Pq$ or $\overline{P}q < Pq$ (or, $\overline{L}q > Lq$ or $\overline{L}q < Lq$) from the above requires that the prices for the high quantity relatives err in the opposite direction as the prices for the low quantity relatives. This condition is likely to be fulfilled only when the prices to be corrected are few. Even then, for government services and capital goods, we have little sure knowledge of the direction in which $\bar{p}(\bar{p}_1 \text{ or } \bar{p}_2)$ differ from p.

If, however, we are convinced that there are some items of government expenditures in which the value of the services does not justify their costs: $p' < 0$, a calculation of the quantity relatives for these particular items may enable us to know the relation between $\overline{P}q$ and Pq as far as the valuation of government services is concerned. This assumes, of course, that it is correct to value the rest of government services at cost.

It also may be inferred that during a period of high prosperity (designated as Situation II) investment is characterized by over-optimism (expectation may be so influential as to turn an inverse price-quantity relationship into a direct one). Thus, prices are too high for the capital goods which rise most in quantity (for increasing prices are expected). To compare this period with a normal one (designated as Situation I), we then have $p'_2 < 0$ for these capital goods of which the quantity relative q_1/q_2 is low (or q_2/q_1 is high), so $\overline{P}q < Pq$.

Perhaps, this device applies better to the case of rationing. Suppose Situation II is under rationing and Situation I is not. The quantity relative q_1/q_2 is likely to be high for most rationed goods for which $p'_2 > 0$, hence that $\overline{P}q < Pq$ seems probable. So, even if $Pq > 1$, we cannot say that real income is higher in Situation II than in Situation I. On the other hand, if $Pq<1$, it is quite apparent that $\overline{P}q<1$ and we should probably say that real income is no higher in Situation Ⅱ than in Situation Ⅰ.①

Here, a word may also be said about the effect of a change in tastes on real income, although confidence in our real income comparison diminishes if tastes are allowed to vary. Suppose, other things being equal, individuals in Situation Ⅱ demonstrate a greater demand for commodity z at the given price, then z should be valued lower than its equilibrium price in Situation Ⅱ in order to ensure that real income is higher in Situation Ⅱ than in Situation Ⅰ if $Pq>1$. However, if q_1/q_2 is high for z, Pq as weighted by the equilibrium price would be satisfactory. This is apparent because Pq will be still greater if z is weighted by a value lower than its equilibrium price.

5.7 The Magnitude by which Virtual Prices Differ from Actual Prices

If our knowledge of the direction in which $\overline{p}$ differs from p does not enable us to decide whether $\overline{P}q$ (or $\overline{L}q$) is greater or less than unity, we must proceed to find the magnitude by which $\overline{p}$ differs from p.

Again it is apparent that there is little possibility of finding out how much $\overline{p}$ differs from p for government services or capital goods. We shall therefore confine ourselves to the problem created by rationing.

A rough estimation of the virtual price for any rationed good requires a knowledge of the price elasticity of demand. Let p and q represent the price and the quantity of the rationed good just before rationing and Δq the change in the quantity of the rationed good as rationing is introduced. Its virtual price is given approximately by

$$\overline{p}=p+\Delta p=p+\frac{p}{q}\cdot\frac{\Delta q}{E}$$

where E is the price elasticity of demand for the rationed good. In addition, if after rationing the changes in the level of income and in other prices are considerable, we need to calculate the income elasticity of demand and the cross elasticities of demand. This is required because the

---

① For a concrete discussion, see Carter, C. F., Reddaway, W. B. and Stone Richard. The Measurement of Production Movements[M]. Cambridge: Cambridge University Press, 1948: 70-71.

demand for any good is, in theory, a function of income (expenditure) and all prices. ①

It should be noticed that the elasticities to which we refer are not the ones which are calculated for some particular income group, such as for the working class. What we need here are the collective elasticities for the economy as a whole. For expediency, these elasticities may be calculated for a selected income group which is believed to be representative of the whole economy, but it is better to take the weighted average (weighted by the number of persons in each income group) of the elasticities estimated for all income groups. It is important to note that the elasticity of the economy will depend on the distribution of income.

In as much as the estimation of $\bar{p}$, either from family budget data or from time series of market data, is subject to error, $\overline{P}q$ (or $\overline{L}q$) can only be stated in terms of probability within limits. It is safe to accept $\overline{P}q > 1$ only when its lower limit exceeds unity at high probability, and to accept $\overline{P}q < 1$ only when its upper limit is exceeded by unity at high probability.

5.8 Comparison of Inputs

Since it is almost impossible to estimate the so-called virtual prices for government services and capital goods, a decision has to be made either to exclude the government and the investment sectors from our valuation or to abandon consumers' utility as the basis for comparison. If the first alternative is chosen, the discussion thus far in this chapter may be unnecessary; if, however, the second alternative is chosen, what substitute

---

① The complete form of the demand function for the ith good is

$$q^i = f(M, p^1, p^2, \cdots, p^s), i = 1, 2, \cdots, s.$$

where M represents the total expenditure. An increment in the demand function in response to the increments in M and the $p'\!s$ is approximately

$$\Delta q^i = \frac{\partial q^i}{\partial M} \Delta M + \frac{\partial q^i}{\partial p'} \Delta p' + \cdots + \frac{\partial q^i}{\partial p^i} \Delta p^i + \cdots + \frac{\partial q^i}{\partial p^s} \Delta p^s.$$

In terms of elasticity,

$$\Delta q^i = E_m \frac{q^i}{M} \Delta M + E_{p'} \frac{q^i}{p'} \Delta p' + \cdots + E_{p^i} \frac{q^i}{p^i} \Delta p^i + \cdots + E_{p^s} \frac{q^i}{p^s} \Delta p^s.$$

In order to determine Δp^i, we have to know the income elasticity of demand E_m and the cross elasticities of demand E'_p (except E^i_p) in addition to the price elasticity E^i_p.

can be used as a basis for comparison?

It has been shown that productive efficiency or productivity may serve as a basis for our comparison. In comparing productivity between two situations, we ask whether all the possible combinations of goods produced in one situation can also be produced in the other situation and the index number criteria which we set up confirm only the impossibility, not the possibility, of producing in one situation the same amounts of goods as actually produced in the other. However, even such limited information requires the assumption that at market prices, every producer maximizes his revenue at a given cost, thus maximizing also his profit. Such maximization behavior certainly does not apply to the government as a producer in the absence of a price system for government services. Moreover, in an imperfect economy, realization may differ from expectation, and it cannot be said that at the market prices the existing goods and services represent the best combination for the sake of maximizing revenue if such a difference occurs. ① The greater the difference between plans and realization, the further is actual revenue likely to be from the point at which revenue would have been maximized, other things being equal. In terms of our iso-efficiency analysis of two goods, the price line will not necessarily be tangential to the iso-efficiency curve—the more the divergency between plans and realization, the further the price line from being tangential to the iso-efficiency curve.

But, though it is hazardous to say that every producer, including the government, maximizes his profit at market prices, it is nevertheless safe to say that every producer, in making his physical product, minimizes his cost at the factor market prices. If there is a free market for the productive factors, the marginal productivity of each factor will be proportional to its price and for each producer the rate of marginal substitution between any two factors is diminishing at their price ratio. The index number criteria for increase of utility thus apply (except that the prices and the quantities

---

① It may be pointed out that, in the comparison of productive efficiency in last chapter, it was tacitly assumed that production is a timeless process.

refer to the productive factors rather than to the consumers' goods). ①

Let us replace p and q with a and x to represent the prices and the quantities of the productive factors. Then, if $\sum a_2 x_2 > \sum a_2 x_1$, we say that it is impossible for every producer to produce in Situation I the same amounts of goods as he actually produces in Situation II. ② If $\sum a_1 x_1 > \sum a_1 x_2$, we say that it is impossible for every producer to produce in Situation II the same amounts of goods as he actually produces in Situation I.

Again, we cannot demonstrate the possibility of producing in any one situation the same amounts of goods as produced in the other situation except in the trivial cases where it can be seen that there is more of every good in one situation than in the other, or that there is more of every productive factor in one situation than in the other. We may probably infer that the greater the amount by which $\sum a_2 x_2$ exceeds $\sum a_2 x_1$, and $\sum a_1 x_2$ exceeds $\sum a_1 x_1$, the greater the possibility of producing in Situation II the same amount of every good as produced in Situation I. This possibility is also greater, the more transferable the factors, x_2, from one use to another, and the more variable the quantities of the factors at constant supply prices—the total value of the factors remaining unchanged.

5.9 A More Detailed Analysis

A further inquiry may be made into the possibility of producing in one situation the same amount of every good as produced in the other.

---

① Admittedly, the data available are not complete enough to warrant a comparison of inputs. In comparing the inputs in two different situations, the task of reducing the difference in quality to quantity seems even more difficult than in comparing one outputs. The important thing is that the difference in efficiency must be fully accounted for in preparing the quantity data.

② It is less accurate to say that, if $\sum a_2 x_2 > \sum a_2 x_1$, it is impossible for the economy as a whole to produce in Situation I the same amounts of goods as those produced in Situation II. To say this would imply an assumption that the costs of the productive economy reach a minimum if the cost of every producer in the economy reaches a minimum: nothing can be gained by, say, a redistribution of jobs among the producers.

If this assumption is not made, it may be possible for the producers of Situation I (though impossible for the producers of Situation II) to produce in Situation I (i.e. with the productive factors of Situation I) the same amounts of goods as produced in Situation II because there may be a change in the distribution of jobs among producers between these two situations.

The general demand function for the kth factor $(k=1,2,\cdots,r)$ is

$$x^k = x^k(q^1, q^2, \cdots, q^s; a^1, a^2, \cdots, a^r)$$

or, for short, $x^k = x^k(q;a)$. Suppose, for the moment, the factor supply prices are constant. Then, the cost of production for q_1 facing the factor prices a_1 is $\sum a_1^k x^k(q_1;a_1) = \sum a_1^k x_1^k$; for q_1 facing a_2 is $\sum a_2^k x^k(q_1;a_2)$; for q_2 facing a_2 is $\sum a_2^k x^k(q_2;a_2) = \sum a_2^k x_2^k$, and for q_2 facing a_1 is $\sum a_1^k x^k(q_2;a_1)$. Thus, if $\sum a_1^k x_1^k > \sum a_1^k x^k(q_2;a_1)$ the possibility of producing in Situation I the same amounts of goods as produced in Situation II is granted; if $\sum a_1^k x_1^k < \sum a_1^k x^k(q_2;a_1)$, the possibility is denied. If $\sum a_2^k x_2^k > \sum a_2^k x^k(q_1;a_2)$, the possibility of producing in Situation II the same amounts of goods as produced in Situation I is granted; if $\sum a_2^k x_2^k < \sum a_2^k x^k(q_1;a_2)$, the possibility is denied.

Let $\sum a_1^k x_1^k - \sum a_1^k x^k(q_2;a_1) = \sum a_1^k \Delta x_1^k$ and $\sum a_2^k x_2^k - \sum a_2^k x^k(q_1;a_2) = \sum a_2^k \Delta x_2^k$. ① Were it not for $a_1 \neq a_2$, we should have $\sum a_1^k \Delta x_1^k = -\sum a_2^k \Delta x_2^k$. If the a_1 relative prices are the same as the a_2 relative prices (i.e. the price ratio between any two factors is the same in both situation), $\sum a_1^k \Delta x_1^k$ and $\sum a_2^k \Delta x_2^k$ will be opposite in sign. It would be enough then to know either $\sum a_1^k \Delta x_1^k$ or $\sum a_2^k \Delta x_2^k$. However, if the relative fluctuations of the prices of a_1 and a_2 are different, we have to calculate both. That is, the possibility of producing in Situation II more of every good produced in Situation I does not preclude the possibility of producing in Situation I more of every good produced in Situation II.

---

① By quadratic approximation,

$$\Delta x_1^k = \sum_{i=1}^{s} \left[\frac{\partial x^k(q_2;a_1)}{\partial q^i}(q^i_2 - q^i_1) + \frac{1}{2}\frac{\partial^2 x^k(q_2;a_1)}{(\partial q^i)^2}(q^i_2 - q^i_1)^2 \right]$$

and

$$\Delta x_2^k = \sum_{i=1}^{s} \left[\frac{\partial x^k(q_1;a_2)}{\partial q^i}(q^i_1 - q^i_2) + \frac{1}{2}\frac{\partial^2 x^k(q_1;a_2)}{(\partial q^i)^2}(q^i_1 - q^i_2)^2 \right].$$

We may hope that a fair notion of the factor demand function can be obtained from the producers' records so that Δx_1^k and Δx_2^k are calculable. See, for example, the argument in Court L M, Lewis H G. Production Cost Indices[J]. Review of Economic Studies, 1942—1943, 10(1): 31.

If there is a distinction between the factor demand functions of the producers of Situation I and those of the producers of Situation II, the possibility or impossibility of producing in one situation the same amount of goods as produced in the other should be referred to the producers whose demand functions are used.

Why does the one possibility not preclude the other if distribution and tastes are the same in both situations? The answer is simple. Inasmuch as government services and capital goods are involved, in neither situation would the combination of goods necessarily represent an optimum from the welfare standpoint. Thus, even if it is possible, say, to produce in Situation II more of every good produced in Situation I, it does not follow that the goods of Situation II are more desirable (again from the welfare standpoint) than the goods of Situation I; even if it does, the possibility of producing in Situation I more of every good produced in Situation II is still not ruled out because the actual combination of goods produced in Situation I may not be the best that can be produced.

Since $\sum a_1^k x^k(q_2;a_1)$ represent a minimum of costs in producing q_2 at a_1 prices and $\sum a_2^k x^k(q_1;a_2)$ represent a minimum of costs in producing q_1 at a_2 prices, we have $\sum a_1^k x_2^k > \sum a_1^k x^k(q_2;a_1)$ and $\sum a_2^k x_1^k > \sum a_2^k x^k(q_1;a_2)$. ①Thus, if $\sum a_2^k x_2^k > \sum a_2^k x_1^k$, $\sum a_2^k x_2^k > \sum a_2^k x^k(q_1;a_2)$; if $\sum a_1^k x_2^k > \sum a_1^k x_1^k$, $\sum a_1^k x_1^k \gtreqless \sum a_1^k x^k(q_2;a_1)$. The Paasche's and the Laspeyres' quantity indexes once more serve as double criteria.

It should not be forgotten that the possibility of producing in one situation the same amount of goods as produced in the other has been granted(or denied) on the assumption that the quantities of the factors can be varied to any extent at constant supply prices in either situation. This assumption actually gives us the most favorable condition of granting such a possibility. We should, in contrast, examine this possibility under the most rigid condition in which the quantities of the factors are assumed as fixed in both situations. That is, is it possible to produce q_1 with x_2 or q_2 with x_1?

In using the factors of one situation to produce the goods of another the following circumstances need be known: first, what is the most efficient distribution of the factors among the producers of different goods, i.e., how much of each factor should be allotted to the making of each

---

① It should be recognized that $\sum x^k(q_2;a_1)$ and $\sum x^k(q_1;a_2)$ here may be written as $\bar{x}_2$ and $\bar{x}_1$ so as to correspond to the hypothetical bundles of goods, $\bar{q}_2$ and $\bar{q}_1$, introduced in the utility analysis.

good. Second, how much of each good can be produced in accordance with such a distribution. This is a general equilibrium problem which cannot be solved without a knowledge of the production functions for all goods together with the amount of expenditure pre-determined for each good. [1]

5.10 Government Interference and Monopolistic Practices

The government may control prices by regulations or it may affect prices by imposing indirect business taxes. Monopolists may raise prices by restricting output. If demand is satisfied at these prices, prices will be proportional to marginal utilities as far as consumers' goods are concerned. A distinction should be made between government floor prices and government ceiling prices as affecting utility index number criteria. When government floor prices are set, buyers may buy to the full extent of their demand though sellers may not be able to sell as much as they wish. When government ceiling prices are set, the reverse is true. Therefore, from the consumer's standpoint, while government floor prices do not disturb the proportionality between prices and marginal utilities, government ceiling prices do. However, it may be pointed out that, though the artificial barriers may not invalidate the utility index number criteria if the proportionality between prices and marginal utilities is not disturbed, they do affect the quantities purchased (or sold) and hence the quantity indexes themselves.

On the other hand, these administered prices will affect the index numbers as criteria for increase of productivity. The goods and services then produced will not represent a maximization of revenue from the economy's standpoint. The proportionality between prices and marginal costs is violated. We cannot say, however, that the quantity index weighted by the prices excluding indirect taxes and monopolistic profits will give us a true measure of productivity. For once indirect taxes and monopolistic elements are removed from the economy, goods and services would be produced in different quantities. If the elasticity of demand for the taxed

---

[1] See Appendix B to this chapter.

and monopolized goods is greater than zero, these goods will be produced in greater quantities. ①

5.11 Restoration of Equilibrium

If we expect to obtain positive knowledge for purposes of making real income comparisons, the prices and the quantities used in making index numbers must be in equilibrium (in the sense specified above). If, however, this equilibrium is disturbed by imperfect elements in the economy, recourse may be had to the following:

(1) Assuming that the critical value of the index numbers, such as $Pq > 1$, would not be affected if the imperfect elements were removed from both the situations compared.

(2) Correcting the value of the index numbers with the estimated equilibrium prices and quantities.

(3) Looking for a comparison for which the prices and the quantities available are in equilibrium.

In the first case, we feel that a rigorous demonstration of the validity of the assumption is lacking; in the last two, the necessary data may not be available.

6. Summary and Conclusion

In a one-commodity world, it is easily demonstrated that real income will be higher in one situation than in another if it is assumed that economic welfare varies directly with the quantity of goods involved. Even in a world with a plurality of commodities, there is no difficulty in telling in which situation real income is higher provided that all commodities increase or decrease in the same proportion. In the real world, however, there is an overwhelming presumption that, while some commodities will increase in amount, others will decrease. As a result, real income comparisons cannot be based on the quantity of goods alone. Knowledge of the economic importance of the goods is also required. Since economic importance is not

---

① Haberler G, Hagen E E. Taxes, Government Expenditures and National Income[J]. Conference on Research in Income and Wealth, 1946, 13:14.

fully measured by price(because of consumer's or producer's surplus), real income comparison must rest on the presumption of the prevalence of optimum conditions in each situation under comparison. Index numbers as indicators of change in real income can provide limited information under restricted circumstances.

First, consider the behavior of the individual consumer. Let us presume, in each situation under comparison, that for the amount of utility obtained the bundle of goods actually bought minimizes the expense. Under these circumstances, it is impossible to obtain the same amount of utility by purchasing any other bundle of goods without incurring additional expense. If conditions are such that the goods of Situation Ⅰ cost as much as or less than the goods of Situation Ⅱ at the prices of Situation Ⅱ, or technically speaking, if Paasche's quantity index is equal to or greater than unity, it can be inferred that with given tastes real income is higher in Situation Ⅱ than in Situation Ⅰ. Conversely, if Laspeyres' quantity index is equal to or less than unity, real income is said to be higher in Situation Ⅰ than in Situation Ⅱ.

Due to the ordinal nature of utility, it is not possible to say how much higher real income is in one situation than in another. Further, in case Paasche's quantity index is less than unity while Laspeyres' quantity index is greater than unity, we cannot tell, in general, in which situation real income is higher.

Second, consider the behavior of a group of individuals insofar as consumption goods are concerned. If income distribution remains unchanged, what has been said for the individual can also be said for the group as a whole. If, however, there is a change in income distribution, it is not satisfactory to say without qualification that real national income is higher in one situation than in another. Interpretations of changes in real national income under these circumstances can be made either from the standpoint of the consumer or from the standpoint of the producer.

From the consumer's standpoint, if Paasche's aggregate quantity index is equal to or greater than unity, it is certain that real national income is higher in Situation Ⅱ than in Situation Ⅰ for Distribution Ⅱ. Whether real national income is also higher in Situation II than in Situation I for Distribution I or for any other possible distribution depends further on the magnitudes of (1) the difference of the Paasche's aggregate quantity index from unity, (2) the change in relative prices, (3) the change in income distribution, (4) the difference in tastes among different individuals and (5) the departure from constant marginal proportion of expenditure. Conversely, if Laspeyres' aggregate quantity index is equal to or less than unity, real national income is higher in Situation I than in Situation II for Distribution I. The change in income distribution makes it possible for real national income to be higher in Situation II than in Situation I for Distribution II and at the same time to be higher in Situation I than in Situation II for Distribution I.

From the producer's standpoint, we may presume that the production of the actual combination of goods maximizes revenue. Thus, it is impossible to produce any other combination of goods which yields the same or greater revenue than is produced by the actual combination of goods. If conditions are such that the goods of Situation II are worth as much as or more than the goods of Situation I at the prices of Situation I, i. e. if Laspeyres' quantity index is equal to or greater than unity, it may be inferred, assuming increasing marginal rate of substitution for products over the relevant range of production, that the same quantities of goods as produced in Situation II cannot be produced in Situation I. Insofar as the production of the actual combination of the goods of Situation II is attempted, productivity is higher in Situation II than in Situation I. If Paasche's quantity index is equal to or less than unity, the converse is true.

Last, as government services and capital goods are evaluated and as the pricing of goods in restricted markets is considered, it becomes more and more difficult to say that utility or any other single magnitude is maximized with the actual quantities of goods and services produced in the actual price (or valuation) system. In order to preserve the validity of the index number criteria, the actual prices must be corrected in such a fashion that utility or some other magnitude becomes maximized in the corrected price system. If such corrections are not feasible, a comparison of inputs—assuming that there is a free market for productive factors and that there are adequate data available—may provide another interpretation regarding the change in real national income. By analogy between individual's output and individual's utility, if Paasche's quantity index for productive factors is equal to or greater than unity, it is impossible for every producer to produce in Situation Ⅰ the same quantities of goods as he actually produces in Situation Ⅱ; and conversely.

Aggregate quantity indexes do not provide equally desirable information for real national income comparison between any two situations. In view of the fact that structures of the economy as to government function, perfection of market, freedom of choice, income distribution and so forth are more nearly alike for one pair of situations than for another and that assumptions regarding similarity of tastes, variety of goods, etc. are more nearly satisfied in one pair of situations than in another, aggregate index numbers usually serve better for intertemporal than for interspatial comparisons, and better for comparisons between near periods than between distant periods. Whether the index number criteria—for example, both Paasche's and Laspeyres' aggregate quantity indexes are greater (or less) than unity—are sufficient for indicating an increase (or decrease) of real national income from one situation to another will depend on the definition of increase (or decrease) of real national income proposed for the particular circumstances under which comparison is to be made.

Finally, it should be stressed that economic welfare, as considered in this thesis, is limited to the part of welfare "that can be brought directly and indirectly into relation with the measuring rod of money."① As differentiated lucidly and precisely by Pigou, economic welfare excludes the non-economic welfare associated with the means by which the goods are produced and the purpose for which they are put to use. Thus, on the one hand, not only the circumstances under which goods are produced are ignored, but the exhaustion of natural resources and human capital is not

---

① Pigou A G. Economics of Welfare[M]. 4 ed. London: MacMillan, 1932: 11.

taken into account. On the other hand, no distinction is made between the use of goods which exert an elevating influence and those which exert a degrading influence on society at large. Consequently, economic welfare interpreted in terms of productivity is not an input-output ratio. Economic welfare interpreted in terms of utility is no barometer of total welfare.

7. Appendix A: A Note on Fisher's Concept of Real Income[①]

Fisher's concept of real income is that, "Income consists of services." By services he means, not the services rendered by a productive factor to the making of a commodity, but the services yielded by a commodity to its consumers. In other words, services are consumers' use of commodities or consumers' utility. A machine, a dwelling and even a loaf of bread are not income; they are capital——a source of income. Income arises only when services are rendered. However, services are distinctive from commodities which yield services. Thus, income may fall even though the amount of commodities increases and income may rise even though the amount of commodities falls provided that capital exists and changes appropriately. In other words, income is measured by the amount of consumption irrespective of capital formation.

Fisher denounces the attempt to measure income while maintaining a given level of capital. The "net product" approach, he maintains, gives no clue to the magnitude of income. What is then measured is not the "actual" income, but rather the "standard" or "regular" income. In such a "standard" income account, not only the value of income but also the value of capital are entered. Though capital may be regarded as a source of future income, its valuation is not on the same basis as current income. The value of income is exactly the value of the actual services, whereas the value of capital in any instant is derived from the value of the future services which that capital is expected to yield. But the expected services may not be the actual services. If the value of the actual services differs

---

① Fisher Iriving. The Nature of Capital and Income[M]. New York: MacMillan, 1906: 104-118, 124, 188, 264.

from the value of the expected services, the "standard" income account would give us a wrong measure of income(even if there is no objection to including the future income in the present income account). Only when expectation is fully realized, is the present value of capital exactly the same as the value of the actual future services discounted at the prevailing rates of interest. The income-value of capital should be nothing but the value of the future services which that capital actually yields.

According to Fisher, what we ought to do is to enter the changeable, fluctuating capital value in a capital account, so that the income account (which records only the value of actual service) reflects correctly the state of welfare.

However, Fisher's income account gives no measure of the income-productivity in each period: what proportion of the actual services received in the present period is produced in the past period and what proportion of the actual services in the future period is produced in the present period? To obtain such a measure, as we desire in this chapter, the present income which has origin in past capital must be recorded(with discount)in the past income account and future income which has origin in present capital must be entered in the present income account. This transfer recording of income is equivalent to entering, in each period, the income value of capital in the "actual" income account. However, the income-value of capital is given not by its present value but by the discounted value of its actual services in the future. This discounted value per unit of capital is what we call the "virtual" price of capital.

In explaining the meaning of the "virtual" price of capital(which may be termed as the "income" price of capital), Fisher's concept of real income is fundamental.

8. Appendix B

The possibility of Producing the Goods of One Situation with the Productive Factors of Another: Producing q_1 with x_2.

Let the production functions for the s goods be written as

$$q^1 = q^1(x^{11}, x^{21}, \cdots, x^{r1}),$$
$$q^2 = q^2(x^{12}, x^{22}, \cdots, x^{r2}),$$
$$\cdots\cdots$$
$$q^s = q^s(x^{1s}, x^{2s}, \cdots, x^{rs}).$$
(1)

Our problem is to find out whether it is possible to produce q_1 with the productive factors given by

$$x^{11} + x^{12} + \cdots + x^{1s} = x_2^1,$$
$$x^{21} + x^{22} + \cdots + x^{2s} = x_2^2,$$
$$\cdots\cdots$$
$$x^{r1} + x^{r2} + \cdots + x^{rs} = x_2^r.$$
(2)

Before we can ascertain the most economical use of the productive factors, we have to assign an amount of expenditure for each good. Thus, if $c^1, c^2, \cdots, c^s$ denote the amounts of expenditure for $q^1, q^2, \cdots, q^s$ respectively, then

$$a^1 x^{11} + a^2 x^{21} + \cdots + a^r x^{r1} = c^1,$$
$$a^1 x^{12} + a^2 x^{22} + \cdots + a^r x^{r2} = c^2,$$
$$\cdots\cdots$$
$$a^1 x^{1s} + a^2 x^{2s} + \cdots + a^r x^{rs} = c^s.$$
(3)

Then, the most economical use of the productive factors is conditioned by

$$\frac{q_{x^{11}}^1}{a^1} = \frac{q_{x^{21}}^1}{a^2} = \cdots = \frac{q_{x^{n1}}^1}{a^n},$$
$$\frac{q_{x^{12}}^2}{a^1} = \frac{q_{x^{22}}^2}{a^2} = \cdots = \frac{q_{x^{n2}}^2}{a^n},$$
$$\cdots\cdots$$
$$\frac{q_{x^{1s}}^s}{a^1} = \frac{q_{x^{2s}}^s}{a^2} = \cdots = \frac{q_{x^{ns}}^s}{a^n}.$$
(4)

where $q_x^i ki$ is the partial derivative of q^i with respect to x^k.

Assuming that the production functions are known and the amount of expenditure for each good is given, we have $r(s+1)$ unknowns:

$$a^1, a^2, \cdots, a^r;$$
$$x^{11}, x^{21}, \cdots, x^{r1};$$
$$x^{12}, x^{22}, \cdots, x^{r2};$$

$$\cdots\cdots$$
$$x^{1s}, x^{2s}, \cdots, x^{rs}.$$

There are also $r(s+1)$ equations r in (2), s in (3) and $(r-1)s$ in (4) which are just enough to determine the unknowns.

But, what values would we give to the $c^1 s$ in (3)? If $c^1, c^2, \cdots, c^s$ are simply given by the costs of production in Situation II, it is possible that while some goods produced here mark an increase, others record a decrease. As the relative factor prices change, there is no assurance that the same proportions of expenditure assigned to the various goods will make the quantities of all goods increase or decrease together.

Acknowledgement

The writer wishes to express his gratitude to Professor Donald W. Paden, chairman of the thesis committee, who gave generously of his time and advice and did much to improve the clarity and significance of the study. The writer is also indebted to Professor Robert Ferber, who read all drafts of the thesis and suggested many important alterations in its content and form. Appreciation is also due Professors J. F. Bell, J. F. Due, C. A. Hickman and J. L. McConnell, who also read different sections of the thesis and made valuable comments.

VITA

Shao Kung Lin was born on December 4, 1922 in Peiping, China. He was graduated in 1944 from National Central University, Chungking, China. Since then, his experience has been in teaching elementary statistical methods and in government statistics. He came to the United States in 1947 and began his graduate work at Louisiana State University in that year. He entered the Graduate College of the University of Illinois September 1949.

第二部分
数理统计

"数理统计是一门估计(推测)的技术,又是一门信息和决策的学科。"

——林少宫

20世纪50年代,国内出现"俄语热",为尽快把概率统计引入国内,林少宫教授自学俄语,成效斐然。1957年,在为华中工学院(现华中科技大学)第一次科学讨论会撰写的论文《从任意两位值间的概率估计论分布函数的确定》中,林教授多次引用俄文文献。1958年,林教授所在的教研室鼓励他写一本数理统计方面的专著,当时国内对统计学的认识是相当模糊的。林教授突出从样本到总体的推断,于1960年写成书稿,并于1963年由人民教育出版社出版了《基础概率与数理统计》一书,填补了国内该学科的空白。该书为许多工科院校所采用,至1980年已第2版第7次印刷,成为后来出版的许多数理统计教科书的编写模式样板。1961年,林教授完成了《信息论讲义》的写作,因出版社审稿能力的原因而未予出版,但在原华中工学院领先讲授。1962年,林教授应邀完成《序贯分析与应用》一文,分别在中国科学院数学所和全国第一次数理统计学术会议做特约学术报告。1979年,林教授应国家教委之约,在大连工学院做概率与统计示范教学。1982年,林教授又受高等教育出版社之托,翻译出版了《应用概率》上、下两册。林教授的这些开创性的工作扫除了统计理论发展中的众多盲点,受其影响,传统的统计学开始重视推断统计,并强调抽样的随机性和概率论的应用。

林教授撰写过多篇数理统计方面的论文,被本章收录的有:1973年在《华中工学院学报(创刊号)》发表的"估计缺落数据的交互对比法",1980年在《统计与管理》上发表的"怎样在加速寿命试验中构造置信区间"和1984年为《工科数学(创刊号)》撰写的"近代数理统计的精巧性和应用性"。

估计缺落数据的交互对比法[①]

提要：实验工作者时而发现某些实验数据缺落而需要另行估计。通常用最小二乘法。本文提出交互对比法，并通过符号法获得了任意个水平的任意个因素的交互对比的简洁形式（以及相应的估值公式），得以统一处理析因实验一类的缺数问题。其次还建立了交互对比法和最小二乘法的等效性。通过实际计算，表明交互对比法具有简单、直接、准确、灵活等特点。

1. 前言

一个实验工作者发现他的个别实验数据缺落，这是常有的事。例如在粉末冶金中，某个压件在烧结过程中发生变形，致使实验落空；又如在汽车轮胎的实地耐磨试验中，偶然发生轮胎破裂，而得不到应有的实验数据等等。由于特殊情况的出现，改变了实验条件，或由于客观条件所限，未能对实验作完整的安排，其结果对实验分析来说，也都无异于缺数情形。

遇到缺数的情形，如果不重复做实验，则一般需要对缺数进行估计，否则将不能对实验结果作出正确的分析。估计缺掉数值的方法，通常是按实验设计的不同类型，使用不同的估值公式。尽管这些公式大多是以最小二乘法（有时是最大似然法）为理论根据，但由于设计类型的繁多，公式的形式不一，既难理解，又不便于记忆，特别是缺少灵活运用的可能性，故本文提出交互对比法，作为统一处理各类缺数问题的一种尝试，并在"部分析因实验"中具体分析和具体应用。所谓交互对比法，就是把"交互作用"和"对比"这两个概念结合起来运用的结果。现分节叙述于后。

[①] 原文发表于《华中工学院学报》1973年12月创刊号。

2. 多个因素水平的复杂性

我们将借助于析因实验中的交互作用的分析和表达，以解决按任何标志划分的实验缺数问题。

我们知道，n 个因素 A、B、C、D、E、$\cdots$，各二个水平的因素效应及各级交互作用可用符号法表达为

$$A = \frac{1}{2^{n-1}} \overbrace{(a-1)(b+1)(c+1)(d+1)(e+1)\cdots}^{n \text{ 个因子}}$$

$$AB = \frac{1}{2^{n-1}} (a-1)(b-1)(c+1)(d+1)(e+1)\cdots \quad (2\text{-}1)$$

$$\cdots\cdots$$

$$ABCDE = \frac{1}{2^{n-1}} (a-1)(b-1)(c-1)(d-1)(e-1)\cdots$$

以上等式右端展开后的各项代表相应因素水平组合的实验结果。例如

$$\text{三因素交互作用 } ABC = \frac{1}{4}[(abc-ab-bc+b)-(ac-a-c+1)] \quad (2\text{-}2)$$

其中，ab 代表 A、B 为高水平，C 为低水平的组合实验结果，等等。但随着因素水平数的增加，通常关于交互作用的分析和表达将变得十分复杂。

例如二个因素各三个水平的情形，共有 $3^2=9$ 种不同的组合。记因素 A 的三个等距水平为 a^0, a^1, a^2；B 的三个等距水平为 b^0, b^1, b^2，则 9 种组合为：

$$\begin{array}{ccc|ccc}
a^0b^0 & a^0b^1 & a^0b^2 & 1 & b & b^2 \\
a^1b^0 & a^1b^1 & a^1b^2 \text{ 或} & a & ab & ab^2 \\
a^2b^0 & a^2b^1 & a^2b^2 & a^2 & a^2b & a^2b^2
\end{array} \quad (\text{按代数中指数运算规则})$$

对于在 a^0 水平上的三个数量（实验结果）a^0b^0, a^0b^1 和 a^0b^2，可分解出 B 的一次作用 B_{L0} 和二次作用 B_{Q0}：

$$B_{L0} = \frac{1}{2}[(a^0b^2 - a^0b^1) + (a^0b^1 - a^0b^0)]$$

$$= \frac{1}{2}(a^0b^2 - a^0b^0) = \frac{1}{2}(b^2 - 1)$$

$$B_{Q0} = \frac{1}{2}[(a^0b^2 - a^0b^1) - (a^0b^1 - a^0b^0)]$$

$$= \frac{1}{2}(a^0 b^2 - 2a^0 b^1 + a^0 b^0)$$

$$= \frac{1}{2}(b^2 - 2b + 1) = \frac{1}{2}(b-1)^2$$

同理,在水平 a^1 上可分解出 B 的一次和二次作用 B_{L1} 和 B_{Q1};在水平 a^2 上可分解出 B_{L2} 和 B_{Q2}。那末,关于 A 和 B 的交互作用,就可以按照同样的分解方法,在 B_{L0},B_{L1},B_{L2} 这三个数的基础上计算

$$\frac{1}{2}(B_{L2} - B_{L0}) \quad \text{和} \quad \frac{1}{2}(B_{L2} - 2B_{L1} + B_{L0})$$

作为 A 对 B 的一次作用的一次影响(称为 A 一次 B 一次交互作用并记为 $A_L B_L$)和 A 对 B 的一次作用的二次影响(称为 A 二次 B 一次交互作用并记为 $A_Q B_L$)。还可以在 B_{Q0}、B_{Q1}、B_{Q2} 这三个数的基础上计算

$$\frac{1}{2}(B_{Q2} - B_{Q0}) \quad \text{和} \quad \frac{1}{2}(B_{Q2} - 2B_{Q1} + B_{Q0})$$

作为 A 对 B 的二次作用的一次影响(称为 A 一次 B 二次交互作用并记为 $A_L B_Q$)和 A 对 B 的二次作用的二次影响(称为 A 二次 B 二次交互作用并记为 $A_Q B_Q$)。这些交互作用是容易进一步展开的,例如 A 二次 B 二次交互作用为

$$A_Q B_Q = \frac{1}{2}(B_{Q2} - 2B_{Q1} + B_{Q0})$$

$$= \frac{1}{4}(a^2 b^2 - 2a^2 b^1 + a^2 b^0 - 2a^1 b^2 + 4a^1 b^1 - 2a^1 b^0 + a^0 b^2 - 2a^0 b^1 + a^0 b^0) \tag{2-3}$$

并且从符号上有

$$A_Q B_Q = \frac{1}{4}(a^2 b^2 - 2a^2 b + a^2 - 2ab^2 + 4ab - 2a + b^2 - 2b + 1)$$

$$= \frac{1}{4}(a^2 - 2a + 1)(b^2 - 2b + 1) = \frac{1}{4}(a-1)^2 (b-1)^2$$

虽然 A 和 B 的各次交互作用都可以从符号上得到较简洁的形式,但随着因素及因素水平的增加,不仅各种交互作用繁多,且其物理意义愈难于区分,故使用价值也愈少。为了解决缺数问题,我们一方面看到高级(指多因素)高次交相作用的分析和表达至为复杂,而另一方面,理论分析和实际经验都表明高级高次交互作用一般趋近于微小。为了利用这后一特点作为插补方法的依据,问题就在于如何得到一种统一的形式,以概括高级高次交互作用。为此我们考虑对比的方法。

3. 对比、交互对比

先从实例出发。

比较两种制造方法的好坏,往往用一种方法的产值 y_1 减另一种方法的产值 y_2,即以其产值差额

$$y_1 - y_2 \tag{3-1}$$

作为衡量。注意这里两个产值的系数是:$+1$ 和 -1。

比较 s 种制造方法的好坏,可以按制造方法的性质,简单地划分为 m 种和 $n-m$ 种相比较($m < n \leqslant s$),并且对不同的 m 和 n 可以进行多次比较。不妨假定取前 m 种和相继的后 $n-m$ 种相比较,并假定这 s 种方法的产值依次为

$$y_1, y_2, \cdots, y_n, y_{n+1}, \cdots, y_s$$

比较的方法将是

$$(n-m)(y_1 + \cdots + y_m) - m(y_{m+1} + \cdots + y_n) + 0(y_{n+1} \cdots y_s) \tag{3-2}$$

注意这 n 个产值的系数是

$$\underbrace{n-m, n-m, \cdots, n-m}_{m\text{个}}; \underbrace{-m, -m, \cdots, -m}_{n-m\text{个}}; \underbrace{0, 0, \cdots, 0}_{s-n\text{个}}$$

如果以两种制造方法的产值比较作为计算单位,则可除以 $(n-m)m$,于是式(3-2)化为

$$\frac{y_1 + y_2 + \cdots + y_m}{m} - \frac{y_{m+1} + \cdots + y_n}{n-m} \tag{3-3}$$

也就是两部分制造方法的平均产值的差额。

在式(3-2)中,当 $m=1, n=s$,即其中一种制造方法同其余制造方法相比时,比较式子为

$$(n-1)y_1 - (y_2 + \cdots + y_n) \tag{3-4}$$

我们注意到,比较式(3-1)到式(3-4)中的产值 y_i 的系数有一个共同的特点,即每个式子中的全部系数之和为零。或者,更一般地说,这些系数符合所谓

对比定义,表达式

$$l_1 y_1 + l_2 y_2 + \cdots + l_n y_n$$

是诸 $y_i (i=1, 2, \cdots, n)$ 的一个对比。如果系数 l_i 满足条件

$$l_1 + l_2 + \cdots + l_n = 0 \tag{3-5}$$

则式(3-4)是 y_1 和 $y_2, \cdots, y_n$ 之间的一个对比。

前节中的因素主效应和因素之间的交互作用如式(3-2)、式(3-3)也都

是实验结果的对比。例如 2^3 析因实验的三因素交互作用 ABC 为

$$[abc-ab-ac-bc+a+b+c-(1)]\div 4$$

其中 8 个实验结果的系数依次是(略去除数 4)

$$1,-1,-1,-1,+1,+1,+1,-1$$

它们显然满足条件式(3-5),又如 3^2 析因实验中两个因素各二次的交互作用 $A_Q B_Q$ 的系数

$$1,-2,1,-2,4,-2,1,-2,1$$

也满足条件式(3-5)。

构成对比的系数看来是十分任意的,但问题不在于对比的形式,而主要在于作出有意义的对比。

我们的目的是为了得到一种含有交互作用意义的对比。首先考虑二个因素多个水平的情形。为此,我们把对比的形成分为三个步骤:

(1) 对比分割

(2) 加权

(3) 交互对比

当实验数据按一个标志(如按某一因素的水平)划分时,可将数据排成一个行矢:

$$y_1, y_2, \cdots, y_n$$

当实验数据按二个标志(如按二因素的水平组合)划分时,可将数据排成一个矩阵:

$$\begin{matrix} y_{11} & y_{12} & \cdots & y_{1n} \\ y_{21} & y_{22} & \cdots & y_{2n} \\ \vdots & \vdots & & \vdots \\ y_{m1} & y_{m2} & \cdots & y_{mn} \end{matrix}$$

当实验数据按多个标志划分时,可将全部数据想象为一个矩形体或超体。对于一个标志的情形,对比的形成无需再加叙述。需要阐述的是二个或多于二个标志的情形。

先考虑二个标志的情形。第一步将矩阵分割如下:

$$\begin{array}{cc|cc} y_{11} & \cdots y_{1k} & y_{1,k+1} & \cdots y_{1n} \\ \vdots & \vdots & \vdots & \vdots \\ y_{h1} & \cdots y_{hk} & y_{h,k+1} & \cdots y_{hn} \\ \hline y_{h+1,1} & \cdots y_{h+1,k} & y_{h+1,k+1} & \cdots y_{h+1,n} \\ \vdots & \vdots & \vdots & \vdots \\ y_{m1} & \cdots y_{mk} & y_{m,k+1} & \cdots y_{mn} \end{array} \quad \left\{\begin{array}{l} 1 \leqslant h < m \\ 1 \leqslant k < n \end{array}\right\}$$

这里各分割区组的元素个数为：

$$hk \qquad h(n-k)$$
$$(m-h)k \qquad (m-h)(n-k)$$

因而，第二步，按区组包含的元素个数加权，配以和它呈反比的系数（绝对值）：

$$\frac{hk(m-h)(n-k)}{hk} \qquad \frac{hk(m-h)(n-k)}{h(n-k)}$$
$$\frac{hk(m-h)(n-k)}{(m-h)k} \qquad \frac{hk(m-h)(n-k)}{(m-h)(n-k)}$$

即

$$(m-h)(n-k) \qquad k(m-h)$$
$$h(n-k) \qquad hk$$

第三步，每个区组的元素总和乘以相应的系数后，按交叉相减的方式作对比[参照式(2-1)]即

$$(m-h)(n-k)(y_{11}+\cdots+y_{hk}) - h(n-k)(y_{h+1,1}+\cdots+y_{mk})$$
$$-k(m-h)(y_{1,k+1}+\cdots+y_{hn}) + hk(y_{h+1,k+1}+\cdots+y_{mn}) \qquad (3-6)$$

我们把对比形式(3-6)叫做交互对比。

对我们特别有意义的分割是其中一个区组只包含一个元素，用它来代表缺数 x。不妨碍一般性，可假设这个元素是 y_{11}（否则可以调换行列来达到这个目的），把它写为 x，于是得到特殊分割

$$\begin{array}{c|ccc} x & y_{12} & \cdots & y_{1n} \\ \hline y_{21} & y_{22} & \cdots & y_{2n} \\ \vdots & \vdots & & \vdots \\ y_{m1} & y_{m2} & \cdots & y_{mn} \end{array}$$

及其相应的交互对比

$$(m-1)(n-1)x - (n-1)(y_{21}+\cdots+y_{m1})$$
$$-(m-1)(y_{12}+\cdots+y_{1n}) + (y_{22}+y_{32}+y_{23}+\cdots+y_{mn}) \qquad (3-7)$$

我们把式(3-7)叫作关于 x 的二级（相当于二因素）交互对比，并用它来概括并代替各次交互作用的划分，以使缺数问题有一个简单的统一解决方法。

不难验证，当 $m=n=3$ 时，式(3-7)中 y_{ij} 的系数就是前面讲的二因素各二次的交互作用式(2-3)中的系数。也就是说，这时式(3-7)相当于一个 4 次交互作用。

我们曾经考虑，高级高次交互作用一般地说是微小的，因此我们将以

此交互对比为零作为条件,从而由式(3-7)得到缺数 x 的估值公式①：

$$x=\frac{(n-1)(y_{21}+\cdots+y_{m1})+(m-1)(y_{12}+\cdots+y_{1n})-(y_{22}+\cdots+y_{mn})}{(m-1)(n-1)}$$
(3-8)

缺数为 $y_{ij}=x$ 时,公式(3-8)可相应地变为：

$$x=\frac{1}{(m-1)(n-1)}\left[(n-1)\sum_{\substack{n=1\\h\neq i}}^{m}y_{hj}+(m-1)\sum_{\substack{k=1\\k\neq j}}^{n}y_{ik}-\sum_{\substack{h=1\\h\neq i}}^{m}\sum_{\substack{k=1\\k\neq j}}^{n}y_{hk}\right]$$
(3-9)

4. 交互对比的一般形式

按三个标志划分的实验结果,可按行、列、层排成一个矩形体。例如三个因素各 m,n,p 个水平的析因实验,可将实验结果排成 m 行,n 列,p 层。用 y_{ijk} 代表第一个因素为第 i 水平、第二个因素为第 j 水平、第三个因素为第 k 水平的因素组合的实验结果,则相应地排在第 i 行第 j 列第 k 层。不妨碍一般性,假定缺数是 $y_{111}=x$。关于 x 的三级交互对比的推导,可借助于一个简单的几何模型。以前后分行,行数为 m；以左右分列,列数为 n；以上下分层,层数为 p。把每个实验结果看作一个元素。将矩形体分割为8个区组,并将区组内的元素总和记为①,②,③,④,⑤,⑥,⑦,⑧,如图1所示。

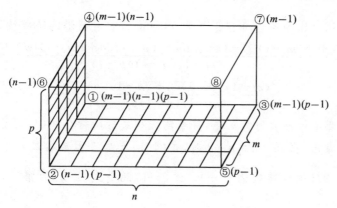

图 1

---

① 但应指出,忽略高级高次作用,如同用低次方程去逼近函数那样,在实际应用中,还应注意奇异点的出现。参考本文,并比较文献：Draper N R, Stoneman D M. Estimating missing values in unreplicated two-level factorial and fractional factorial designs[J]. Biometrics,1964,20(3). 的讨论。

其中：

① 代表缺数 x，
② 代表与 x 同列同层但不同行的元素总和，
③ 代表与 x 同行同层但不同列的元素总和，
④ 代表与 x 同行同列但不同层的元素总和，
⑤ 代表与 x 同层但不同行和列的元素总和，
⑥ 代表与 x 同列但不同行和层的元素总和，
⑦ 代表与 x 同行但不同列和层的元素总和，
⑧ 代表与 x 不同行、列、层的元素总和。

按区组包含元素的个数分别对①、②、③、④、⑤、⑥、⑦、⑧加权（见图1），并使正负号符合二水平的三因素交互作用概念，即得关于 x 的三级交互对比（略去常数因子 $\frac{1}{4}$）：

$$[⑧-(n-1)⑥-(m-1)⑦+(m-1)(n-1)④]$$
$$-[(p-1)⑤-(n-1)(p-1)②-(m-1)(p-1)③$$
$$+(m-1)(n-1)(p-1)①]$$
$$=⑧-(m-1)⑦-(n-1)⑥-(p-1)⑤+(m-1)(n-1)④$$
$$+(m-1)(p-1)③+(n-1)(p-1)②-(m-1)(n-1)(p-1)①$$
$$\tag{4-1}$$

利用零交互对比为条件并把①写为 x，即得三因素情形的缺数估计公式：

$$x=\frac{⑧-(m-1)⑦-(n-1)⑥-(p-1)⑤+(m-1)(n-1)④}{(m-1)(n-1)(p-1)}$$
$$+\frac{(m-1)(p-1)③+(n-1)(p-1)②}{(m-1)(n-1)(p-1)} \tag{4-2}$$

即使 x 在第 i 行第 j 列第 k 层，符号意义和最后所得公式(4-2)都不变。

容易看出，当 $m=n=p=2$ 时，式(4-1)就是三因素各二水平的三级交互作用(1-2)（忽略常数因子 $\frac{1}{4}$）。式(4-1)既是式(2-2)的推广，我们就试图把符号

①，②，③，④，⑤，⑥，⑦，⑧

改写为

abc, bc, ac, ab, c, b, a, (1)

并从符号上推出如下形式：

$$⑧-(m-1)⑦-(n-1)⑥-(p-1)⑤+(m-1)(n-1)④$$

$$+(m-1)(p-1)③+(n-1)(p-1)②-(m-1)(n-1)(p-1)①$$
$$=(1)-(m-1)a-(n-1)b-(p-1)c+(m-1)(n-1)ab$$
$$+(m-1)(p-1)ac+(n-1)(p-1)bc-(m-1)(n-1)(p-1)abc$$
$$=[1-(m-1)a][1-(n-1)b][1-(p-1)b]$$
(4-3)

从而估值公式(4-2)可写为
$$[1-(m-1)a][1-(n-1)b][1-(p-1)c]=0 \qquad (4-4)$$

在符号表达式(4-3)中取 $p=1$ 作为特例,即得二因素情形的交互对比:
$$[1-(m-1)a][1-(n-1)b] \qquad (4-5)$$

以及相应的估值公式:
$$[1-(m-1)a][1-(n-1)b]=0 \qquad (4-6)$$

还容易将式(4-3)推广到任意多个因素、任意多个水平的情形。例如 4 个因素各 m,n,p,q 个水平的交互对比为
$$[1-(m-1)a][1-(n-1)b][1-(p-1)c][1-(q-1)d] \qquad (4-7)$$

及其相应的估值公式:
$$[1-(m-1)a][1-(n-1)b][1-(p-1)c][1-(q-1)d]=0 \quad (4-8)$$

其中(将符号表达式展开后)

$abcd=x$,即代表与 x 有完全相同的因素水平的实验数据,

abd 代表仅与 x 有相同的第一、第二、第四因素水平的实验数据总和,

ab 代表仅与 x 有相同的第一、第二因素水平的实验数据总和,

c 代表仅与 x 有相同的第三因素水平的数据总和,

……

因素的个数再多,估值公式的形式也完全是类似的;根据前面对比分割的方法以及各级交互作用的含义,就能推知,每增多一个因素,公式中就多一个用括号括起来的因式,每个因式的形式完全一样。由此我们得到估计缺数的一般公式:
$$\prod_i [1-(n_i-1)f_i]=0 \qquad (4-9)$$

其中 $\prod$ 为乘积符号,n_i 代表第 i 个因素 F_i 的水平数。表达式展开后的符号 $f_1 f_2 f_3 \cdots$ 具有类似于 $abc\cdots$ 的意义。

式(4-9)有着十分简洁的形式,不仅可用来解决任意多个因素任意多个水平的缺数问题,而且可用来解决按任何标志划分的实验缺数问题,例

如对于随机区组的实验安排,可把区组的划分看作一个因素;对于重复实验的情形,又可把实验的重复次序看作另一个因素等等。

当缺数不止一个时,令缺数为 $x_1, x_2, \cdots$,一般可逐个地对 $x_i(i=1,2,\cdots)$ 应用公式(4-9),从而得到一组线性方程,其方程的个数等于缺数的个数,但缺整个一行(或列)的情形除外。

例:按磷(P)、二硫化钼(M)、烧结温度(T)三个标志划分,重复实验(R)二次的汽车离合器衬套压溃压力如下表:

	T_1		T_2	
	M_1	M_2	M_1	M_2
P_1	587,600	620,642	562,553	x_1,575
P_2	570,x_2	605,638	632,598	632,633
P_3	580,585	600,605	572,574	586,574

表中每格左边的数据代表第一次实验结果,右边的为第二次结果。把重复次序 R 看作第四个因素,利用一般公式(4-9),取 $n_1=3, n_2=2, n_3=2, n_4=2$。关于缺数 x_1(相当于 $P_1 M_2 T_2 R_1$,其中 R_1 代表第一次重复)有

$$\prod_{i=1}^{4} [1-(n_i-1)f_i]$$
$$= (1) - (n_1-1)f_1 - (n_2-1)f_2 - (n_3-1)f_3 - (n_4-1)f_4$$
$$+ (n_1-1)(n_2-1)f_1 f_2 + (n_1-1)(n_3-1)f_1 f_3$$
$$+ (n_1-1)(n_4-1)f_1 f_4 + (n_2-1)(n_3-1)f_2 f_3$$
$$+ (n_2-1)(n_4-1)f_2 f_4 + (n_3-1)(n_4-1)f_3 f_4$$
$$- (n_1-1)(n_2-1)(n_3-1)f_1 f_2 f_3$$
$$- (n_1-1)(n_2-1)(n_4-1)f_1 f_2 f_4$$
$$- (n_1-1)(n_3-1)(n_4-1)f_1 f_3 f_4$$
$$- (n_2-1)(n_3-1)(n_4-1)f_2 f_3 f_4$$
$$+ (n_1-1)(n_2-1)(n_3-1)(n_4-1)f_1 f_2 f_3 f_4$$
$$= (x_2+585) - 2(600) - (638+605) - (598+574)$$
$$- (570+580) + 2(642) + 2(553) + 2(587)$$
$$+ (633+574) + (605+600) + (632+572)$$
$$- 2(575) - 2(620) - (562) - (632+586) + 2x_1 = 0,$$

即
$$2x_1 + x_2 = 1732$$

同理,关于 x_2 有
$$x_1 + 2x_2 = 1697$$

解联立方程得缺落数据的估计值：
$$\begin{cases} x_1 = 589 & （实验值＝577）\\ x_2 = 554 & （实验值＝560）\end{cases}$$

还容易验算，若只缺 x_1，则估值为 586；若只缺 x_2，则估计值与实验值无异。此外，还容易察知，当缺数不止一个且皆属于同一 P_i 水平时，估值公式将不适用。

5. 最小二乘法

我们知道，在多因素的方差分析中，总平方和可表为各级作用及交互作用的平方和。例如考虑 α 个因素，各 $t_1, \cdots, t_\alpha$ 个水平。记 $y(j_1, \cdots, j_\alpha; a_{j_1}, \cdots, a_{j_\alpha})$ 为第 j_1 因素第 a_{j_1} 水平，第 j_2 因素第 a_{j_2} 水平，$\cdots$，第 j_α 因素第 a_{j_α} 水平的实验结果；$I(k_1, \cdots, k_\beta; a_{k_1}, \cdots, a_{k_\beta})$ 为第 $k_1, \cdots, k_\beta$ 诸因素关于水平 $a_{k_1}, \cdots, a_{k_\beta}$ 的 β 级交互作用。我们有[①]

$$\sum_{a_1} \cdots \sum_{a_\alpha} [y(1, \cdots, \alpha; a_1, \cdots, a_\alpha)]^2$$
$$= \sum_{\beta=0}^{\alpha} \sum_{1, \cdots, \alpha}^{k_1, \cdots, k_\beta} \frac{t_1 \cdots t_\alpha}{t_{k_1} \cdots t_{k_\beta}} \sum_{a_{k_1}} \cdots \sum_{a_{k_\beta}} [I(k_1, \cdots, k_\beta; a_{k_1}, \cdots, a_{k_\beta})]^2 \tag{5-1}$$

其中，符号 $\sum_{1, \cdots, \alpha}^{k_1, \cdots, k_\beta}$ 表示对选自 $1, \cdots, \alpha$ 的所有组合 $k_1, \cdots, k_\beta (k_1 < k_2 < \cdots < k_\beta)$ 求和，并且当 $\beta = 0$ 时，取 $t_{k_1} \cdots t_{k_\beta} = 1$，$I =$ 全部实验数据的平均数。另一方面，k 级交互作用平方和 Q_k 又可表为各级平方和的交错总和[②]：

$$Q_k \equiv \sum_{a_{i_1}} \cdots \sum_{a_{i_k}} [I(i_1, \cdots, i_k; a_{i_1}, \cdots, a_{i_k})]^2$$
$$= \sum_{r=0}^{k} (-1)^{k-r} \sum_{i_1, \cdots, i_k}^{j_1, \cdots, j_r} \sum_{a_{j_1}} \cdots \sum_{a_{j_r}} \frac{t_{j_1} \cdots t_{j_r}}{t_{i_1} \cdots t_{i_k}} [y(j_1, \cdots, j_r; a_{j_1}, \cdots, a_{j_r})]^2 \tag{5-2}$$

下面我们证明

定理：交互对比为零与交互作用平方和的极小化等效。也就是，无碍于一般性，令缺数

$$x = y(1, \cdots, k; 1, \cdots, 1)$$

---

① 证明见 Mann H B. 试验的分析与设计[M]. 张里千，等译. 北京：科学出版社，1964. 这里所用符号及其含义均略有改变。

② 证明见 Mann H B. 试验的分析与设计[M]. 张里千，等译. 北京：科学出版社，1964. 这里所用符号及其含义均略有改变。

则由条件
$$\frac{\partial Q_k}{\partial x} = 0$$
推出交互对比法的一般公式
$$\prod_{\nu=1}^{k}[1-(t_\nu-1)f_\nu]=0$$

证:由(5-2),
$$Q_k \equiv \sum_{a_1}\cdots\sum_{a_k}[I(1,\cdots,k;a_1,\cdots,a_k)]^2$$
$$= \sum_{r=0}^{k}(-1)^{k-r}\sum_{1,\cdots,k}^{j_1,\cdots,j_r}\sum_{a_{j_1}}\cdots\sum_{a_{j_r}}\frac{t_{j_1}\cdots t_{j_r}}{t_1\cdots t_k}[y(j_1,\cdots,j_r;a_{j_1}\cdots a_{j_r})]^2$$
$$= \sum_{r=0}^{k}(-1)^{k-r}\sum_{1,\cdots,k}^{j_1,\cdots,j_r}\frac{t_{j_1}\cdots t_{j_r}}{t_1\cdots t_k}[y(j_1,\cdots,j_r;1,\cdots,1)]^2$$
$$+ \sum_{r=0}^{k}(-1)^{k-r}\sum_{1,\cdots,k}^{j_1,\cdots,j_r}\sum_{Ua_{j_1}\neq 1,\cdots,a_{j_r}\neq 1}\cdots\cdots\sum\frac{t_{j_1}\cdots t_{j_r}}{t_1\cdots t_k}[y(j_1,\cdots,j_r;a_{j_1},\cdots,a_{j_r})]^2$$

其中,$Ua_{j_1}\neq 1,\cdots,a_{j_r}\neq 1$ 为逻辑和符号,表示至少有一个 $a_{j_i}\neq 1$。

令 $\frac{\partial Q_k}{\partial x}=0$,于是得
$$\sum_{r=0}^{k}(-1)^{k-r}\sum_{1,\cdots,k}^{j_1,\cdots,j_r}\frac{t_{j_1}\cdots t_{j_r}}{t_1\cdots t_k}[y(j_1,\cdots,j_r;1,\cdots,1)]=0$$

记
$$L \equiv \sum_{r=0}^{k}(-1)^{k-r}\sum_{1,\cdots,k}^{j_1,\cdots,j_r}\frac{t_{j_1}\cdots t_{j_r}}{t_1\cdots t_k}[y(j_1,\cdots,j_r;1,\cdots,1)]$$
$$= \sum_{r=0}^{k}\sum_{1,\cdots,k}^{j_1,\cdots,j_r}(-1)^{k-r}\frac{t_{j_1}\cdots t_{j_r}}{t_1\cdots t_k}[y(j_1,\cdots,j_r;1,\cdots,1)] \tag{5-3}$$

由于
$$y(j_1,\cdots,j_r;1,\cdots,1)$$
$$= \sum_{a_{j_{r+1}}}\cdots\sum_{a_{j_k}}y(j_1,\cdots,j_r,j_{r+1},\cdots,j_k;1,\cdots,1,a_{j_{r+1}},\cdots,a_{j_k})$$
$$= \sum_{n=0}^{k-r}\sum_{(a_{j_{r+1}},\cdots,a_{j_k})_n=1}\cdots\cdots\sum y(j_1,\cdots,j_r,j_{r+1},\cdots,j_k;1,\cdots,1,a_{j_{r+1}},\cdots,a_{j_k})$$

这里用符号 $(a_{j_{r+1}},\cdots,a_{j_k})_n=1$ 表示 $a_{j_{r+1}},\cdots,a_{j_k}$ 中的 n 个 $=1$ 其余 $\neq 1$,作为求和的限制条件,故对于固定的 $j_1,\cdots,j_r$,
$$\sum_{(a_{j_{r+1}},\cdots,a_{j_k})_0=1}\cdots\sum y(j_1,\cdots,j_r,j_{r+1},\cdots,j_k;1,\cdots,1,a_{j_{r+1}},\cdots,a_{j_k})$$

的系数 C 为

$$C = (-1)^{k-r} \frac{t_{j_1}\cdots t_{j_r}}{t_1\cdots t_k} + (-1)^{k-(r-1)} \sum_{t_{j_1},\cdots,t_{j_r}}^{t'_{j_1},\cdots,t'_{j_{r-1}}} \frac{t'_{j_1}\cdots t'_{j_{r-1}}}{t_1\cdots t_k}$$

$$+ (-1)^{k-(r-2)} \sum_{t_{j_1},\cdots,t_{j_r}}^{t''_{j_1},\cdots,t''_{j_{r-2}}} \frac{t''_{j_1}\cdots t''_{j_{r-2}}}{t_1\cdots t_k} + \cdots$$

$$+ (-1)^{k-1} \sum_{t_{j_1},\cdots,t_{j_r}}^{t_{j_i}} \frac{t_{j_i}^{(r-1)}}{t_1\cdots t_k} + (-1)^k \frac{1}{t_1\cdots t_k}$$

由恒等式
$$\prod_{r=1}^{k}(n_r - 1) \equiv \sum_{r=0}^{k}(-1)^{k-r}\sum_{1,\cdots,k}^{j_1,\cdots,j_r} n_{j_1}\cdots n_{j_r} \qquad (5\text{-}4)$$

即知

$$C = (-1)^{k-r}\frac{1}{t_1\cdots t_k}(t_{j_1}-1)(t_{j_2}-1)\cdots(t_{j_r}-1)$$

因而由(5-3)(注意 $\sum_{(a_{j_{r+1}},\cdots,a_{j_k})_0=1}\cdots\cdots\sum$ 已归并到 $\sum_{(a_{j_{r+1}},\cdots,a_{j_k})_0=1}\cdots\cdots\sum$),

$$L = \sum_{r=0}^{k}\sum_{1,\cdots,k}^{j_1,\cdots,j_r}(-1)^{k-r}\frac{1}{t_1\cdots t_k}(t_{j_1}-1)(t_{j_2}-1)\cdots(t_{j_r}-1)$$
$$\cdot \sum_{(a_{j_{r+1}},\cdots,a_{j_k})_0=1}\cdots\cdots\sum y(j_1,\cdots,j_r,j_{r+1},\cdots,j_k;1,\cdots,1,a_{j_{r+1}},\cdots,a_{j_k})$$

若将

$$\sum_{(a_{j_{r+1}},\cdots,a_{j_k})_0=1}\cdots\cdots\sum y(j_1,\cdots,j_r,j_{r+1},\cdots,j_k;1,\cdots,1,a_{j_{r+1}},\cdots,a_{j_k})$$

写为 $f_{j_1}\cdots f_{j_r}$,表示仅与缺数 x 有相同的第 $j_1,\cdots,j_r$ 因素水平的实验数据总和,再注意到恒等式(5-4),便得到

$$L = \sum_{r=0}^{k}\sum_{1,\cdots,k}^{j_1,\cdots,j_r}(-1)^{k-r}\frac{1}{t_1\cdots t_k}$$
$$\cdot (t_{j_1}-1)(t_{j_2}-1)\cdots(t_{j_r}-1)f_{j_1}f_{j_2}\cdots f_{j_r}$$
$$= \frac{1}{t_1\cdots t_k}\prod_{\nu=1}^{k}[(t_\nu-1)f_\nu - 1]$$

故由 $L=0$ 得
$$\prod_{\nu=1}^{k}[1-(t_\nu-1)f_\nu] = 0$$

证毕。

以代表行、列、层的三个因素各 $t_1=m, t_2=n, t_3=p$ 个水平为例,

$$Q_3 \equiv \sum_{a_1}\sum_{a_2}\sum_{a_3}[I(1,2,3;a_1,a_2,a_3)]^2$$

$$= \sum_{a_1}\sum_{a_2}\sum_{a_3}[y(1,2,3;a_1,a_2,a_3)]^2 - \sum_{a_2}\sum_{a_3}\frac{[y(2,3;a_2,a_3)]^2}{t_1}$$

$$- \sum_{a_1}\sum_{a_3}\frac{[y(1,3;a_1,a_3)]^2}{t_2} - \sum_{a_1}\sum_{a_2}\frac{[y(1,2;a_1,a_2)]^2}{t_3}$$

$$+ \sum_{a_3}\frac{[y(3;a_3)]^2}{t_1 t_2} + \sum_{a_2}\frac{[y(2;a_2)]^2}{t_1 t_3} + \sum_{a_1}\frac{[y(1;a_1)]^2}{t_2 t_3} - \frac{y^2}{t_1 t_2 t_3}$$

$$= \sum_i\sum_j\sum_k y_{ijk}^2 - \frac{\sum_j\sum_k y_{\cdot jk}^2}{m} - \frac{\sum_j\sum_k y_{i\cdot k}^2}{n} - \frac{\sum_i\sum_j y_{ij\cdot}^2}{p}$$

$$+ \frac{\sum_k y_{\cdot\cdot k}^2}{mn} + \frac{\sum_j y_{\cdot j\cdot}^2}{mp} + \frac{\sum_i y_{i\cdot\cdot}^2}{np} - \frac{y_{\cdot\cdot\cdot}^2}{mnp}$$

$$= x^2 - \frac{(y'_{\cdot 11}+x)^2}{m} - \frac{(y'_{1\cdot 1}+x)^2}{n} - \frac{(y'_{11\cdot}+x)^2}{p}$$

$$+ \frac{(y'_{\cdot\cdot 1}+x)^2}{mn} + \frac{(y'_{\cdot 1\cdot}+x)^2}{mp} + \frac{(y'_{1\cdot\cdot}+x)^2}{np} - \frac{(y'_{\cdot\cdot\cdot}+x)^2}{mnp}$$

$$+ 不含\ x\ 的项,$$

其中,$y'_{1\cdot\cdot}=$与 x 同行但不包含 x 的数据总和,

$y'_{1\cdot\cdot}+x=$第 1 行数据总和(指各层的第一行相加),

$y_{i\cdot\cdot}=$第 i 行数据总和,

$y'_{\cdot 1\cdot}=$与 x 同列但不包含 x 的数据总和,

$y_{\cdot j\cdot}=$第 j 列数据总和,

$y'_{\cdot\cdot 1}=$与 x 同层但不包含 x 的数据总和,

$y_{\cdot\cdot k}=$第 k 层数据总和,

$y'_{11\cdot}=$与 x 同行同列但不包含 x 的(各层)数据总和,

$y'_{11\cdot}+x=$第 1 行第 1 列数据总和(指各层的位于第 1 行第 1 列的数相加),

$y_{ij\cdot}=$第 i 行第 j 列(各层)数据总和,

$y'_{1\cdot 1}=$与 x 同行同层但不包含 x 的(各列)数据总和,

$y_{i\cdot k}=$第 i 行第 k 层(各列)数据总和,

$y'_{\cdot 11}=$与 x 同列同层但不包含 x 的(各行)数据总和,

$y_{\cdot jk}=$第 j 列第 k 层(各行)数据总和,

$y'_{\cdot\cdot\cdot}=$除 x 外的全部数据总和。

令 $\dfrac{\partial Q_3}{\partial x}=0$ 得

$$x - \dfrac{y'_{\cdot11}+x}{m} - \dfrac{y'_{1\cdot1}+x}{n} - \dfrac{y'_{11\cdot}+x}{p}$$
$$+ \dfrac{y'_{\cdot\cdot1}+x}{mn} + \dfrac{y'_{\cdot1\cdot}+x}{mp} + \dfrac{y'_{1\cdot\cdot}+x}{np} - \dfrac{y'_{\cdots}+x}{mnp} = 0,$$

化简并整理得

$$(m-1)(n-1)(p-1)x = y'_{\cdots} - my'_{1\cdot\cdot} - ny'_{\cdot1\cdot} - py'_{\cdot\cdot1}$$
$$+ mny'_{11\cdot} + mpy'_{1\cdot1} + npy'_{\cdot11} \quad (5\text{-}5)$$

同本节的几何模型对照一下，得知

$y'_{\cdots} = ⑧+⑦+⑥+⑤+④+③+②$

$y'_{1\cdot\cdot} = ⑦+④+③$

$y'_{\cdot1\cdot} = ⑥+④+②$

$y'_{\cdot\cdot1} = ⑤+③+②$

$y'_{11\cdot} = ④$

$y'_{1\cdot1} = ③$

$y'_{\cdot11} = ②$

故由式(5-5)得

$$(m-1)(n-1)(p-1)x = ⑧ - m⑦ + ⑦ - n⑥ + ⑥ - p⑥ + ⑤$$
$$+ mp④ - m④ - n④ + ④$$
$$+ mp③ - m③ - p③ + ③$$
$$+ np② - n② - p② + ②$$

即

$$(m-1)(n-1)(p-1)x = ⑧ - (m-1)⑦ - (n-1)⑥ - (p-1)⑤$$
$$+ (m-1)(n-1)④ + (m-1)(p-1)③$$
$$+ (n-1)(p-1)②$$

$$(5\text{-}6)$$

等式(5-6)和公式(5-2)完全一致。

容易看出，若缺数为 $y_{ijk}=x$，则(5-5)相应地变为：

$$(m-1)(n-1)(p-1)x = y'_{\cdots} - my'_{i\cdot\cdot} - yp'_{\cdot j\cdot} - py'_{\cdot\cdot k}$$
$$+ mny'_{ij\cdot} + mpy'_{i\cdot k} + npy'_{\cdot jk} \quad (5\text{-}7)$$

这一节证明，每一级的交互对比法都可以找到相应的最小二乘法作为进一步的理论对照。但因素较多时，最小二乘法的推导，随着交互作用项的增多而愈来愈复杂,而交互对比法则直接而简单。

6. 在拉丁方上的应用——内加条件法

我们知道,所谓拉丁方,是在行数等于列数的一个方块上排上个数与行数或列数相等的配方,使得每种配方在每行只出现一次,并且在每列也只出现一次。例如,配方数＝行数＝列数＝4 的一个拉丁方,又称边为 4 的拉丁方是:

$$
\begin{array}{cccc}
A_{1\alpha} & B_{2\beta} & C_{3\gamma} & D_{4\delta} \\
B_{4\gamma} & A_{3\delta} & D_{2\alpha} & C_{1\beta} \\
C_{2\delta} & D_{1\gamma} & A_{4\beta} & B_{3\alpha} \\
D_{3\beta} & C_{4\alpha} & B_{1\delta} & A_{2\gamma}
\end{array}
\tag{6-1}
$$

这里 A、B、C、D 代表 4 种配方(暂时不去管其他符号)。

把行看作一个因素的 4 个水平,把列看作另一因素的 4 个水平,配方又看作第三因素的 4 个水平,拉丁方的安排就相当于三个因素各 4 个水平的 1/4 部分实施,(因为三个因素各 4 个水平共有 4^3 种组合,拉丁方的安排只选做其中的 $4^2=16$ 种。)这种部分实验的安排,实际上假定了第一、二个因素间的某种对比(交互作用)为零。例如在未安排第三因素之前,即假定位于 A、B、C、D 的 4 组二因素实验数据(4 个总和)之间的对比为零,即

$$l_1\sum_1^4 A + l_2\sum_1^4 B + l_3\sum_1^4 C + l_4\sum_1^4 D = 0$$

其中 $\sum l = 0$。符号 $\sum_1^4 A=$ 位于 A 的 4 个数的总和,它的含义是假想在未安排第三因素之前位于 A 的 4 个二因素实验结果的总和。其他符号的含义,照此类推。

设第一行第一列为缺数,记以 x。由于 x 落在 A 的位置,故取对比

$$3\sum_1^4 A - \left(\sum_1^4 B + \sum_1^4 C + \sum_1^4 D\right) = 0$$

或

$$3\left(x+\sum_1^3 A\right) - \left(\sum_1^4 B + \sum_1^4 C + \sum_1^4 D\right) = 0 \tag{6-2}$$

作为我们的交互对比法的一个内加条件。(我们称内加条件是因为它本身不是一个要被满足的独立条件,理由见后。)或再将式(6-2)简写为

$$3(x+T') = G' - T' \tag{6-3}$$

其中: T' 是与缺数 x 有相同配方但不包含 x 的数据总和,

G' 是除 x 外的全部数据总和。

一般，当拉丁方的边为 m 时，式(6-3)变为
$$(m-1)(x+T')=G-T'$$
或
$$(m-1)x=G'-mT' \qquad (6-4)$$

对于边为 m 的行列方阵，我们曾导出估值公式(3-8)，将符号改变一下就可写为
$$(m-1)(m-1)x=mR'+mC'-G' \qquad (6-5)$$
其中，$R'=$ 与 x 同行但不包含 x 的数据总和，
$C'=$ 与 x 同列但不包含 x 的数据总和，
$G'=$ 除 x 外的全部数据总和。

由于实际上按 A、B、C、D 4 个位置安排了第三因素，就不能再认为这 4 个位置之间的对比为零，因此考虑零交互对比这个估计条件时，应把对比式(6-4)从中扣除掉。于是我们从式(6-5)的两端扣除式(6-4)这个等量部分，即从式(6-5)减去式(6-4)而得到：
$$(m-1)(m-1)x-(m-1)x=mR'+mC'-G'-(G'-mT')$$
即
$$(m-1)(m-2)x=m(R'+C'+T')-2G' \qquad (6-6)$$
这就是通常采用的拉丁方缺数估计公式。

部分实施的原理是按正交对比为零的条件继续安排新因素。每安排一个新因素，即须扣除一个正交的对比条件。对拉丁方安排第四、第五因素时，可按正交拉丁方安排。例如对边为 4 的拉丁方安排第四个因素的 4 个水平可用号码 1、2、3、4 表示(对这些数字号码加上一个圈以免误解)：

①　②　③　④
④　③　②　①
②　①　④　③
③　④　①　②

其正交特点是，除了每个号码在每行、每列只出现一次外，并且和每个字母也只相遇一次，见式(6-1)。在 1、2、3、4 的位置安排第四个因素的 4 个水平，又意味着在未安排第四个因素之前，1、2、3、4 之间的三因素实验结果的对比为零。

不妨碍一般性，仍假定第一行第一列为缺数 x，那末，作为第二个内加条件将有
$$3\sum_1^4 ① - (\sum_1^4 ② + \sum_1^4 ③ + \sum_1^4 ④) = 0$$

或
$$3(x+\sum_1^3 ①)-(\sum_1^4 ②+\sum_1^4 ③+\sum_1^4 ④)=0$$

或简写成
$$3(x+N')=G'-N'$$

其中 N' 是与 x 有相同号码但不包含 x 的实验数据总和。对边为 m 的情形就有

$$(m-1)(x+N')=G'-N'$$

即
$$(m-1)x=G'-mN' \qquad (6\text{-}7)$$

由于实际上安排了第四个因素,故以式(6-7)作为第二个内加条件,再从式(6-6)中扣除掉,于是得到估值公式:

$$(m-1)(m-3)x=m(R'+C'+T'+N')-3G' \qquad (6\text{-}8)$$

可以证明(从略),式(6-6)和式(6-8)都可以用相应的最小二乘法得到。

内加条件法可以继续推广到加排第五、第六以至第 m 个因素的情形,这时将使用 3,4 直至 $m-2$ 个正交拉丁方,每个拉丁方的使用给原来的交互对比多增加一个有待扣除的正交条件。因为对于边为 m 的拉丁方(如果存在有正交拉丁方的话),至多有 $m-1$ 个正交拉丁方[参考式(6-1)中的希腊字母的安排],故得到如下

命题:对于边为 m 的拉丁方,第 $m-1$ 个内加条件将使估值公式变为恒等式 $0=0$。

论证:既然缺数 x 占第一行,第一列,第一拉丁字母 A,第一数码 ①,第一希腊字母 α,……的位置,那么,按正交性质,除了拉丁字母 A 外,其余的 A 不可能再出现于第一行,第一列;除了一个数码 ① 外,其余的 ① 不可能再出现于第一行,第一列和第一拉丁字母 A 的位置;除了一个希腊字母 α 外,其余的 α 不可能再出现于第一行、第一列、第一拉丁字母 A 和第一数码 ① 的位置;……如此类推。故每个新因素的第一水平除第一次安排在 x 处外,以后只能安排在第一行、第一列、第一拉丁字母、第一数码、第一希腊字母……以外的位置。因此式(6-8)中的 R',C',T',N' 都是位置互不重叠的元素的总和。另一方面,除 x 外,共有 $m^2-1=(m+1)(m-1)$ 个位置。每个因素占去其中的 $m-1$ 个位置(连 x 在内是 m 个),故总共可容纳 $m+1$ 个因素,除行、列两个因素外,还有 $m+1-2=m-1$ 个因素,这也就是正交拉丁方(如果存在的话)的个数。故第 $m-1$ 个内加条件将使式(6-8)退化为

$$(m-1)(m-m)x=m\ (\overbrace{R'+C'+T'+N'+\cdots}^{m-1\ \text{项}})-mG'$$

即 $0=0$。

这个恒等关系正说明内加条件法的合理性。

应用：北京特殊钢厂使用 $L_{16}(4^5)$ 正交表，对三个因素各四个水平安排了 16 次试验，以未安排因素的两列作为误差列。对应的实验结果如下页表（李士琦，1972）。作为以上命题的一个直接应用，是按下面的与最小二乘法等效的方法估计 x_1 和 x_2：

关于 x_1 做对比得
$$3(x_1-2.4)+3(x_1-7.9)-(x_2-13.3-3.1+20.9)$$
$$-(9.1+x_2+23.2-22.3)=0$$
即
$$3x_1-x_2=22.7$$

关于 x_2 做对比得
$$3(x_2-13.3)+3(x_2+23.2)-(-3.1+x_1-2.4+20.9)$$
$$-(9.1+x_1-7.9-22.3)=0$$
即
$$3x_2-x_1=-17.7$$

由此解出 $x_1=6.3, x_2=-3.8$。

	误差列		实验值－38	说明
	1	1	8.7	
	2	2	12.0	
	3	3	x_1	
	4	4	-10.6	
	3	4	-1.6	
	4	3	4.5	
	1	2	x_2	排有因素的三列相当于此表的行、列、拉丁字母（略）。
	2	1	-14.6	
	4	2	19.5	误差列相当于此表的希腊字母和数码。
	3	1	7.5	x_1, x_2 为缺落实验值。
	2	4	-12.5	
	1	3	-24.4	表下端Ⅰ,Ⅱ,Ⅲ,Ⅳ栏内的数值分别为表中四个水平 1,2,3,4 所对应的实验值的总和，例如，$x_2-13.3=8.7+x_2-24.4+2.4$ $-3.1=12.0-14.6-12.5+12.0$ 等等。
	2	3	12.0	
	1	4	2.4	
	4	1	7.5	
	3	2	-8.3	
Ⅰ	$x_2-13.3$	91		
Ⅱ	-3.1	$x_2+23.2$		
Ⅲ	$x_1-2.4$	$x_1-7.9$		
Ⅳ	20.9	-22.3		

作为部分析因实验的拉丁方和正交拉丁方的安排,是以二因素的全面安排为基础的多因素部分实施。至于以三个或多个因素的全面安排为基础的部分实施,有些可比为一种拉丁体或超体①。虽然原则上未尝不可以根据内加条件的设想,对缺数进行估计,但由于交互作用与主效应的互相混杂,不易得出唯一易行的方法。

7. 在部分析因实验上的应用

用"内加条件法"估计拉丁方的缺数问题的现实性,在于假定了因素之间不存在任何交互作用。而对于拉丁体和超体的实验安排来说,一般只假定某些多因素交互作用不存在,而另一些交互作用则仍须考虑。由于部分实施中因素作用互相混杂,如何选取交互对比并令其为零,从而估计缺数,将有一定的选择性。也正是因为这样,交互对比法在这个问题上表现出一定的灵活性。

具体情况,具体分析。现通过实例说明如下。

北京大学数力系为做橡胶补强剂的冲击弹性试验,利用正交表 $L_{27}(3^{13})$ 安排了一个 6 因素各 3 个水平的部分实施。实验安排及结果摘录如下页表。

现在我们假设第 1 号试验数据缺落(即缺掉 13 这个数)并记以 x,那么,在Ⅰ,Ⅱ,Ⅲ栏内的数只能登记为:

	A	B	C	D	E	F
Ⅰ	$59+x$	$-20+x$	$-24+x$	$16+x$	$4+x$	$-26+x$
Ⅱ	-25	-20	-8	-11	-25	-9
Ⅲ	-78	-4	-12	-49	-23	-9

因为实验的安排是在 3 个因素各 3 个水平的全面实验(即 $3^3=27$ 个实验)的基础上进行的,所以我们准备选出作用(或效应)最为显著的 3 个因素,对它们使用交互对比法(这里要求选出的 3 个因素组成一个 3^3 全面试验)。这样做的理由是,我们预期这些因素的交互对比的大小最少受其他因素作用的干扰。Ⅰ,Ⅱ,Ⅲ栏中的数是对应于每一因素的每一水平的实验数据总和,从这些总和数值看,对应于 A、D、E 这三个因素的数值变化幅度较大:A 的变化幅度很可能是从 -78 到 $59+x$(虽然我们不知道 x 是什

---

① 关于拉丁体和正交拉丁体定义,可参考:Kishen K. On the construction of Latin and Hyper-Graeco-Latin cubes and hypercubes[J]. J Indian Soc Agr Stat,1949:22-23.

么值),D 的变化幅度很可能为 -49 到 $16+x$,E 的变化幅度很可能为 -23 到 $4+x$,而对应于其他因素的数值变化幅度则较小。变化幅度的大小反映因素作用的大小,因此我们选择 A、D、E 这三个因素来对它们使用交互对比法。

做出这一选择之后,就回到三因素情形的估计公式(4-4)。注意只考虑 A、D、E 三列因素水平,对应于第 1 号试验的水平是 $(1,1,1)$。把 A、D、E 三个因素依次地看作第一、第二、第三因素。于是将公式(4-4)展开并取 $m=n=p=3$ 得

$$x=\frac{1}{8}[(1)-2a-2b-2c+4ab+4ac+4bc]$$

$$=\frac{1}{8}[(1)-2(a+b+c)+4(ab+ac+bc)]$$

其中 $(1)=$ 与 x 有完全不同因素水平的实验数据总和 $=-5-8-5-10-17-13-10-12=-80$,

$a=$ 仅与 x 有相同的第一因素水平的数据总和 $=8+7+5+4=24$,

$b=$ 仅与 x 有相同的第二因素水平的数据总和 $=0+1-6-5=-10$,

$c=$ 仅与 x 有相同的第三因素水平的数据总和 $=-5-1-14-1=-21$,

$ab=$ 仅与 x 有相同的第一、第二因素水平的数据总和 $=9+9=18$,

$ac=$ 仅与 x 有相同的第一、第三因素水平的数据总和 $=11+6=17$,

$bc=$ 仅与 x 有相同的第二、第三因素水平的数据总和 $=8+0=8$,

$abc=x$,

故

$$x=\frac{1}{8}[(-80)-2(24-10-21)+4(18+17+8)]=\frac{106}{8}=13.25$$

(非常接近实验值 13)

6 因素各 3 个水平的实验安排如表 1 所示。

表 1　6 因素各 3 水平实验安排

试验序号	A	B	C	D	E	F	Y−50
1	1	1	1	1	1	1	13
2	1	1	2	2	2	2	8
3	1	1	3	3	3	3	7
4	1	2	1	2	3	3	5

续表

试验序号	A	B	C	D	E	F	Y−50
5	1	2	2	3	1	1	6
6	1	2	3	1	2	2	9
7	1	3	1	3	2	2	4
8	1	3	2	1	3	3	9
9	1	3	3	2	1	1	11
10	2	1	1	2	2	3	−5
11	2	1	2	3	3	1	−8
12	2	1	3	1	1	2	8
13	2	2	1	3	1	2	−5
14	2	2	2	1	2	3	0
15	2	2	3	2	3	1	−5
16	2	3	1	1	3	1	1
17	2	3	2	2	1	2	−1
18	2	3	3	3	2	3	−10
19	3	1	1	3	3	2	−17
20	3	1	2	1	1	3	0
21	3	1	3	2	2	1	−13
22	3	2	1	1	2	1	−6
23	3	2	2	2	3	2	−10
24	3	2	3	3	1	3	−14
25	3	3	1	2	1	3	−1
26	3	3	2	3	2	1	−12
27	3	3	3	1	3	2	−5
Ⅰ	72	−7	−11	29	17	−13	
Ⅱ	−25	−20	−8	−11	−25	−9	
Ⅲ	−78	−4	−12	−49	−23	−9	
T				−31			

说明：$A、B、C、D、E、F$ 代表 6 个因素，表中 1、2、3 代表它们的水平。Y 代表实验结果(百分数)。

Ⅰ 表示同一列中的 1 所对应的($Y-50$)的总和。例如 C 列的 1 对应的总和是 $13+5+4-5-5+1-17-6-1=-11$，故把这个数对准 C 列登记在Ⅰ栏内。Ⅱ、Ⅲ的意义仿此。T 表示全部($Y-50$)值的总和，即 $13+8+\cdots+(-5)=-31$。

我们再来假设第 14 号实验数据缺落（即缺掉数据 0），对应于这个序号的因素水平为 (2,2,2,1,2,3)，故 Ⅰ、Ⅱ、Ⅲ 栏的数值应登记为：

Ⅰ　　72　　−7　　−11　　29+x　　17　　−13
Ⅱ　　−25+x　−20+x　−8+x　−11　　−25+x　−9
Ⅲ　　−78　　−4　　−12　　−49　　−23　　−9+x

从这些数看来，仍然是对应于 A、D、E 的数值变化幅度较大，因此仍然选 A、D、E 为作用较显著的因素。为了利用 Ⅰ、Ⅱ、Ⅲ 和 T 栏的已有部分总和进行计算，可将公式 (4-4) 转换成公式 (5-7)，似更方便。这里 $m=n=p=3$，故公式进一步化为[注意第 14 号实验中 (A,D,E) 的水平为 $(2,1,2)$]：

$$(m-1)^3 x = Y'_{\cdots} - m(Y'_{i\cdot\cdot} + Y'_{\cdot j\cdot} + Y'_{\cdot\cdot k}) + m^2(Y'_{ij\cdot} + Y'_{i\cdot k} + Y'_{\cdot jk})$$
$$= Y'_{\cdots} - m(Y'_{2\cdot\cdot} + Y'_{\cdot 1\cdot} + Y'_{\cdot\cdot 2}) + m^2(Y'_{21\cdot} + Y'_{2\cdot 2} + Y'_{\cdot 12})$$

将具体数值代入得

$$8x = -31 - 3(-25 + 29 - 25) + 9[(8+1) + (-5-10) + (9-6)] = -8$$
$$x = -1 \quad （比较接近实验值 0）$$

通过以上的实例验算，表明当缺数不多时，这种"选较显著因素"的方法是行得通的，即既有可能选出作用比较显著的因素，而且估值也比较接近实验值。

在本例中，估值之所以接近实验值并非出于偶然。由于 D、E 分别安排在交互作用列 ABC 和 AB^2C 上，故 A、D、E 交互作用 $= A \times ABC \times AB^2C = A^3 B^3 C^2 = C^2$（指数按模数 3 化简），而表中的 C 列作用确实微小。

若 C 列也显著，则可适当考虑使用内加条件法[①]。

8. 结语

本文的基本思想是建立在因素间的高级高次交互作用趋近于微小的理论分析和实践基础之上。先通过"对比"的概念，简化了多水平的因素交互作用，从而导出统一的缺数估计公式，再借助于符号法，使公式进一步化为有明显规则的简洁形式。为了概括重要的"部分析因实验"的缺数问题，提出了"内加条件"的观点和"选较显著因素"的方法。通过实例验算，表明"选较显著因素"法在实验次数较多而缺数较少的情形下往往是行得通的。

---

① 刘璋温同志在审阅本文讨论稿时，曾以 $L_{27}(3^{13})$ 表第 10 列（即 $A^2 BC$ 列）为误差列，按最小二乘法（极小化"误差"方差）插补第 1 号和第 14 号数据为 21.5 和 −0.5（这两个插补值都可以利用简单的对比法求得），表明了在部分实施的补缺问题上存在着一定的选择性与灵活性。若从估值的准确度考虑，则用 C 列比用第 10 列进行插补，效果更好。

鉴于"对比"的广泛含义,还希望能从交互对比法概括更多的实验类型的缺数问题。例如对于 Kempthorne 提出的有区组混杂的补缺方法(Kempthorne O,1952),是可以用交互对比的概念去解释的。但交互对比法如何应用于有区组混杂的情形以及其他类型的实验设计上面,则尚待研究。本文仅就用"对比"概念统一处理各类缺数问题的可能性作一初步探讨,工作做得还很不完善。

交互对比法的使用,并不限于析因实验,可用于按任何标志划分的实验安排。但预料对数量水平的析因实验比按其他标志划分的实验将更为合适。并且预料因素愈多,因素的水平相隔愈近,估值愈可能准确。然而由于实验误差的存在,在实验次数较少的情况下,估值的准确性问题实有待于具体分析;估计值接近实验值并不等于说估计一定是准确的(Draper & Stoneman,1964)。

通过最小二乘法的推导,建立了交互对比法和最小二乘法的等效性。因此使用交互对比法,对常用的 F 概率分布检验的影响,无须作新的叙述。(虽然交互对比法不是从误差的角度考虑缺数的添补问题,但实际上把非零的微小高级交互对比值视为一种误差。)除应用于"部分析因实验"外,交互对比法和最小二乘法的方法不同而结果一致。因此交互对比法可视为对最小二乘法的一种"物理"解释。比较而言,交互对比法是一种直接的"物理"方法;而最小二乘法则是一种间接的数学方法,两者从概念上互相补充。

但是,交互对比法和最小二乘法之间的等效性的建立,还不等于说交互对比法和最小二乘法是一样的。因为最小二乘法还要考虑"误差"方差 Q 由什么构成,例如并不排除在多于二个因素的情形中把某些二级交互作用项包括在 Q 之内。另一方面,按照对比和交互作用的定义,当然也还可以根据具体问题的物理意义考虑不同的对比和交互作用,不一定要使用最高级的交互对比。(正好比用高于二次的多项式去逼近一个本身就是二次式的函数,是完全多余的。)但这只说明交互对比法的灵活性,而不是它的局限性和约束性,同时也说明补缺问题本身就是一个多种多样的问题。

※ ※ ※

承中国科学院计算所徐钟济老师,数学所张里千、刘璋温同志及中国科学技术大学数学系统计运筹室对本文讨论稿提出宝贵意见和建议,得以改进本文的若干处,作者在此表示衷心的感谢。

近代数理统计的精巧性和应用性

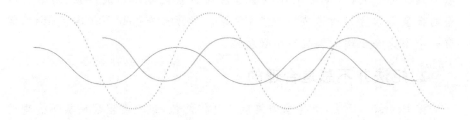

数理统计被喻为:打开几个小窗户就能了解世界。意思是说,通过抽样观测能对总体进行种种推断。

1. 什么是统计推断?

统计推断主要是概率计算;用概率语言描述对总体的了解。

但统计推断与概率问题有别。例如,

概率问题:假定次品率 p,计算 n 次抽样(独立重复试验)有 r 次为次品的概率[这个问题的解为 $C_n^r p^r (1-p)^{n-r}$]。

统计推断:假定 n 次抽样中有 r 次为次品,问次品率情况如何?设 n 次抽样有 r 次为次品的概率为 P,则 P 随次品率 p 不同而不同,从 P 和 p 的对应关系而推测次品率情况。

首先,有一个如何提问次品率情况的问题。传统的提问是:

(1) p 是什么?从而作点估计或区间估计。

(2) $p=p_0$?或 $p \neq p_0$?从而检验假设。由于提问不中肯,往往影响到观测或实验数据的有效利用。A. Wald(1939)提出统计判决的观点,提问实验者真正关心的问题。例如,在工艺的革新试验中,问哪些工艺可望把次品率至少降低到 5%?

其次,要讲究解决问题的方法。半个多世纪以来,统计学家一直在考虑统计推断的准确性问题。K. Pearson 及其学派大多以直观作为统计推断的基础,如作为估计方法的矩法即是。R. Fisher 从 1912 年开始,指出直

---

① 原文发表于《工科数学》(现为《大学数学》),1984(2):5-9.

观方法的效率问题。但直到 1936 年，J. Neyman 和 E. S. Pearson 才首次给出统计推断的精确数学模型，其特点为：(1)对不正确判断给以明显的惩罚，(2)造出惩罚最小的、有效利用数据的统计方法。惩罚的大小，自然要联系到实际损失，如实际平均产量与估计产量之差所造成的损失，把不合格的批量划为合格所造成的损失，等等。尽管损失函数不能准确衡量，但有一个不完全正确的模型要比直观强。

2. 矩法并不总是合理的

按 Pearson 矩法，用样本均值估计总体均值，但也会出现明显不合理的情况。假定观测到某家庭两次生下的婴孩都是男性，问下一次出生婴孩的性别是什么？按矩法就会预测 2/2＝1 即 100% 是男性。这样的估计岂不近于谬误！再看 J. Kiefer 列举的一个例子(Kiefer J,1969)，设质点在某时刻位于 θ。$\theta-1$ 或 $\theta+1$ 的概率各 $1/3$。θ 待估。设三次测量结果为 x_1, x_2, x_3 且无测量误差。如果三个测量值中有二个相同，而另一个又相差 2 个单位，比方说，$x_1=14, x_2=16, x_3=14$，则必然有 $\theta=15$，而样本均值 $=\frac{44}{3}$ 就把 θ 低估了 $1/3$。容易计算，出现这种局面的概率是 $2/9$。对统计模型作准确概率分析，比任何直观方法都更能选出合理的推断方案。

3. 凭什么进行估计？

假定要判断一批水果的次品率，用的抽样方法可比作一个抛硬币的试验模型，为了易于说明问题，假定出现正面 H(代表次品)的概率 p 只有两个可能值 $\frac{1}{3}$ 和 $\frac{2}{3}$，现进行 3 次独立试验，根据试验结果判断 $p=\frac{1}{3}$ 或 $p=\frac{2}{3}$。再者，除非特别声明，不妨认为凡是错误的估计都蒙受同样的损失。且考虑如下的一些估计方案(Kiefer J,1969)：

(1) 若 3 次试验中 H 至少出现 2 次，则估计 $p=\frac{2}{3}$；否则估计 $p=\frac{1}{3}$。

(2) 若 H 比 T(T 表示反面，代表正品)先出现，则估计 $p=\frac{2}{3}$；否则估计 $p=\frac{1}{3}$。

(3) 若 H 至少出现一次，则估计 $p=\frac{2}{3}$；否则估计 $p=\frac{1}{3}$。

(4) 不管试验结果如何，总是估计 $p=\frac{2}{3}$。

共有 $HHH, HHT, \cdots, TTT$ 等 8 种试验结果，分别对 $p=\frac{1}{3}$ 和 $p=\frac{2}{3}$ 算出各种结果的概率后，就可整理出估计正确的概率如下：

方案	当 $p=\frac{1}{3}$ 时 估计正确的概率	当 $p=\frac{2}{3}$ 时 估计正确的概率
1	20/27	20/27
2	18/27	18/27
3	8/27	26/27
4	0	1

例如，当 $p=\frac{1}{3}$ 时，这 8 种结果的概率依次为：

结果	HHH	HHT	HTH	THH	HTT	THT	TTH	TTT
概率	$\frac{1}{27}$	$\frac{2}{27}$	$\frac{2}{27}$	$\frac{2}{27}$	$\frac{4}{27}$	$\frac{4}{27}$	$\frac{4}{27}$	$\frac{8}{27}$

故对方案 1 来说，H 至少出现二次的概率为 $\frac{7}{27}$；H 出现一次或不出现的概率为 $1-\frac{7}{27}=\frac{20}{27}$。

我们说，方案 2 是不容许的，因为方案 2 在任何情况下都比方案 1 差。但怎样在容许的方案 1、3、4 之中进行选择，还可以提出某些准则。如

(1) 给定一种正确概率，求另一种正确概率最大的方案。例如，要求在 $p=\frac{1}{3}$ 时正确概率 $\geqslant\frac{8}{27}$，则方案 3 为最优。Neyman 和 Pearson 准则的精神实质正在这里。问题也许是为什么要选 $\frac{8}{27}$ 作为一种界限，通常要求这个正确的概率高达 95% 或 99%。

(2) 极大极小准则。从最大错误概率中选最小者。方案 1 的最大错误概率 $\max\left\{\frac{7}{27},\frac{7}{27}\right\}=\frac{7}{27}$，是所有方案中的最小者，故方案 1 最优。

(如将试验次数减为 2，则最好的估计方案将是：H 出现 2 次估计 $p=\frac{2}{3}$；T 出现 2 次时估计 $p=\frac{1}{3}$；H、T 出现一次时估计 $p=\frac{1}{3}$ 或 $\frac{2}{3}$ 的机会各半，这是一种随机估计方案。)

(3) 对称准则。因为 H 与 T 交换等于概率 $\frac{1}{3}$ 与 $\frac{2}{3}$ 交换,故若试验结果为 TTH 时估计 $p=\frac{1}{3}$,则试验结果为 HHT 时理应估计 $p=\frac{2}{3}$,如此等等。在容许方案中,仅方案 1 具备此对称性(又称不变性),且可证明,它是对称方案中的最优者。

并非对所有模型都存在容许的对称方案。如果将两个概率 $\frac{1}{3}$ 和 $\frac{2}{3}$ 改为 $\frac{1}{5}$ 和 $\frac{2}{3}$,交换 H 与 T,这两个概率就变为 $\frac{4}{5}$ 和 $\frac{1}{3}$,并不存在对称性。

(4) 加权平均最小准则。例如赋予先验概率

$$P(p=\frac{2}{3})=\frac{3}{4} \text{ 及 } P(p=\frac{1}{3})=\frac{1}{4}$$

按此先验概率对方案 3 加权平均为

$$\frac{3}{4}\times\frac{26}{27}+\frac{1}{4}\times\frac{8}{27}=\frac{43}{54}$$

这比任何其他方案的加权平均都大,故方案 3 最优。(若按 $\frac{9}{10}$ 和 $\frac{1}{10}$ 加权,则方案 5 最优。)

(5) 最小平均损失。除赋予先验概率如上外,再给定损失函数如下表:

	估计 $p=\frac{1}{3}$	估计 $p=\frac{2}{3}$
次品率 $p=\frac{1}{3}$	0	1
次品率 $p=\frac{2}{3}$	2	0

读如次品率 $p=\frac{2}{3}$ 而估计 $p=\frac{1}{3}$ 的损失为 2 单位,等等。估计正确时损失为零,于是容易算出方案 3 蒙受的平均损失为

$$2\times\left(1-\frac{26}{27}\right)\times\frac{3}{4}+1\times\left(1-\frac{8}{27}\right)\times\frac{1}{4}=\frac{25}{108}$$

这比其他方案所蒙受的平均损失要小,而且是相对于所赋予的先验概率和所给定的损失函数的最优方案(称 Bayes 决策方案)。

(事实上,每一个容许方案都可以成为相对于某先验概率或某损失函数的 Bayes 决策方案!)

以上说明,不同准则、不同背景(如先验概率或损失函数),导致不同的

最优方案。在统计推断中,并无简单药方足以对付一切情形,这种令人困惑的局面是任何其他数学领域中的问题都不能比拟的。不提出准则而凭直观行事则更危险。例如统计文献中有不少建议方案 2 的,面对方案 5 全然无视!

现代统计推断理论,就是强调模型,讲求准则,走向优化。从以上讨论中,还看到两个明显的趋向:一是先验知识的利用,二是行动的损失估计。前者复活了历史上几经起落的贝叶斯观点;后者则表明统计判决理论在未来统计学中日益明显的价值。

4. 显著性判断,不是乱猜?

在 Wald(1944)引入序贯分析之前,样本大小总是在做试验前预先规定好的(当然 Dodge 和 Romig 的二次抽样是一个例外)。例如,对于次品率 $p=\frac{1}{5}$ 的假设 H_0,当次品数在抽样中占的比例显著地超过 $\frac{1}{5}$ 时,就认为假设 H_0 不足信。5% 显著性的判据,既可以是抽查 50 件中有 15 件次品;也可以是抽查 100 件中有 27 件次品;等等。总之,样本的大小,50 也好,100 也好,是预先规定的。不同的样本大小,要求有不同的次品数作为显著性判据。所有这些数据都保证不接受一个真实的"假设 $H_0:p=\frac{1}{5}$"的概率 $\leqslant \alpha=5\%$。但是,我们决不能只抽查到发现有显著性判据出现就停止抽查,并就此认为显著性判据已说明 H_0 不足信。这种不按预定样本大小办事而随意停止抽样的方法,不能保证 $\alpha=5\%$,它要比 5% 大得多。重对数律告诉我们它要大到 1(无限多次接近极限边界的概率=1)(Feller William, 1968)!

试验次数	显著性判据	极限边界
30	9	9.4
40	12	12.1
50	15	14.7
70	20	19.7
100	27	27.0
200	50	50.3
300	72	72.9

400	94	95.0
500	115	117.0
1000	221	224.9

我们知道，一个对等的二人博弈，如果任由一方随意停止博弈，那将是不公平的，同理，统计推断可以看成是人们和自然界之间的博弈，如果先是对既定的样本大小算出显著性判据，其后又随便改变样本大小，并企图按原来算好的显著性判据进行判断，也就是"不公平"的了。然而，在统计应用中，任意改变样本大小而不改变显著性判据计算的做法不乏其例。甚至认为只要能解决实际问题，就不必追求严格！竟不知这种做法几近乱猜。

看来，不预先规定样本大小而作显著性判断是个困难问题。必须重新设法计算显著性判据。序贯分析成功地、而且异常简单地解决了这个问题。它不要求预先规定样本大小，仍然保证 $\alpha=5\%$（或其他显著水平）！事实上，一个假设 H_0 总是相对于另一个假设 H_1（比方说，$H_1,p=p_1$）而言，序贯分析，在同时保证不接受一个真实的"$H_0:p=p_0$"的概率$\leq\alpha$ 和接受一个虚伪的"$H_0:p=p_0$"（这时"$H_1:p=p_1$"真实而被拒绝）的概率$\leq\beta$ 的条件下，制定出停止抽样并作结论的规则。一箭双雕！

对于通常给定的 α 和 β，序贯分析往往（平均地说）把试验次数（即样本大小）减少到一半以下（和预先规定的样本大小相比），故对破坏性试验，如武器试验，特别有用。因此它在第二次世界大战期间曾多年被划为绝密资料。序贯分析不但有重大应用价值，而且有深远的理论意义。建立在序贯分析基础上的判决函数的一般理论（Wald,1950），不仅把古典推断方法作为特例，还把实验设计和统计推断（从原始数据的观测到加工分析作出结论）一体化，它提供了最一般的统计理论模型。

Wald 和 Wolfowitz(1948)证明在给定两类错误概率 α 和 β 下，序贯分析中的概率比检验（SPRT）使试验平均次数达到最小，从而把统计学在最优化方面大大推进一步。自此以后，人们不仅考虑推断方案的最优，而且考虑实验方案的最优。

5. Fisher,Wald,各有所执

Wald 把假设检验和参数估计都看作统计决策；用决策的观点解释、利用检验和估计理论，为制定最优行动策略服务。对此，Fisher 大不以为然，他并不赞成在数理统计中考虑最优化应用（Fisher R A,1956）。当然，谁也

不能忽视最优化思想在应用统计中的价值,但问题在于最优行动策略是不是数理统计本身的内容?像我们开头引过的那句话,数理统计学是为了打开几个小窗户就能了解世界。如果考虑最优决策,这门科学就似乎没有适当的内涵和范围了。然而,在本文的多处,我们又看到,怎样有效地了解世界,是和我们的应用目的有关的。

第二部分 数理设计

抽样调查的代表性问题及有关理论和概念的历史发展

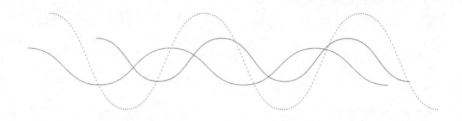

1. 引言

调查样本的设计与估计问题是不可分割的。例如，J. Neyman(1934)曾把有代表性的抽样方法定义为能够对所调查的总体特性作出区间估计的方法。本文从一些历史事实分析抽样设计的代表性概念及其演变，并形成了如下一种看法，即所谓目的性抽样并非纯属历史兴趣，今日的概率样本设计可视为历史上目的性抽样的概率翻版；仅对可能的目的性样本赋予概率意义而已。

为了使叙述上有较大的连续性，以下分设计和估计两个部分讨论。

2. 调查样本设计

2.1 历史的前奏

从部分推算全体，由来已久。例如，据记载英商 J. Graunt(1662)曾对伦敦城内保有较完整登记表册的教区做家庭调查，发现平均每年 11 个家庭举行 3 次葬礼，而伦敦每年共举行葬礼约 13000 次，故得结论：家庭总数约为 48000。令每家庭平均人口为 8 人，便估计得伦敦总人口约 384000 人。虽然 Graunt 知道这些统计均值随时随地而异，但他根据一个教区的洗礼和葬礼次数推算全国的人口数量，表明了他对比率的恒定性的某种信心。17—18 世纪的人口统计学家包括英国的 Wm Petty 和 E. Halley，瑞典的 Per Wargentin 和德国的 J. P. Susmilch，都曾根据一个地区的部分数据资料，对整个地区做过类似的粗糙推算。

P. S. Laplace(1812)在他的"概率的分析理论"一书中,讲到法国政府根据他的请求,做了一次人口统计的抽样实验。"在分布在全国的30个部门里,考虑要抵消气候差异的影响,选择了那些由于社长的热忱和智慧而能够提供最精确信息的公社,进行调查。"

这些早期抽样调查的特点是:首先考虑样本数据资料的完整性和精确性(以防止非抽样误差),然后凭主观愿望(或想当然地)认为总体(就出生率和死亡率而言)是均匀的。因此也就缺乏任何抽样设计的思想。

我国的两句成语:"举一反三","闻一知十",虽然可解释为一种抽样实践,但从抽样设计的观点看,也缺乏明确的涵义。

2.2 代表性样本

20世纪末之前,在统计调查中很少用抽样的方法,少数的抽样调查也是伴随着普查而进行的,人们不认为抽查可以代替普查,因此也没有对抽样设计作认真的努力。及至20世纪末本世纪初,挪威统计学家A. N. Kiaer提倡代表性调查(representative investigation),认为通过有代表性的抽样调查,可以达到了解总体的目的,抽样设计才成了一门学问。Kiaer认为,抽样调查的准确性主要问题不在于样本的大小,而在于样本的代表性。典型(平均情况)调查虽然有用,但弊病甚多。为了得到一个有代表性的样本,使它能成为全国的一个缩影,它必须包含分布在全国各地的许多单元。为此,抽样单元不是随便选取的,必须首先"把全国划分为许多调查区。例如在社会调查中,首先区分城市与乡村,然后再把它们细分为大、中、小;海岸、内地;工业区、农村。要仔细地把一个国家划分为各个均匀的部分,才能获得真正有代表性的样本。"还要求这些划分要符合普查所提供的线索,以便抽样结果能受到控制,统计数据能得到充分分析。

虽然完全随机的抽样方式已在某种程度上得到公认,但Kiaer并不认为随机抽样是唯一的最好方式。

1926年国际统计学会指出,在选取有代表性样本的许多方法中,要区分两种方法:

(1)随机选取。一些单元的选取,要使得其中每一单元都有完全相等的被选中机会。

(2)有目的地选取若干组单元,使得它们合并起来能产生和总体几乎一样的特性。为了能知道估计值的准确性,应包括有足够多组单元,以便能测出各组之间特性的变异。

可以看出,代表性抽样是一种分层抽样,但究竟怎样抽取,仍然是不够明确的,特别是关于目的性的抽取方法,缺乏数学描述。

英国统计学家 A. L. Bowley(1926)除主张曾由 Kiaer 一度提出的按比例分层抽样外,对目的性抽样做了数学理论上的补充;把它描述为一种分群抽样。假定所调查的数量 Y 和若干个叫做控制量的特性有相关关系,并假定 Y 的群均值对每一控制量的群均值的回归都是线性的,那么,群的选择应使每个控制量的群均值都近似地等于它的总体均值。

如上所述,目的性抽样需要对总体有非常详细的了解。然而,如果这种抽样没有任何随机性,就无法应用概率理论。因此,Neyman(1934)明确指出,所谓目的性抽样,不外是分层的分群随机抽样而已,从而排除了专门对目的性抽样另作理论探讨的必要性。

2.3 Neyman 的贡献

Neyman(1934)的论文对抽样调查理论作出了划时代的贡献,他把代表性抽样归结为分层抽样,并规定了层内抽样必须随机化,包括按群随机。这不但使一度含糊的代表性抽样这个历史上的概念得到澄清,而且明确表示,"分层的分群随机抽样是唯一可以推荐的通用抽样方法"。但他认为一般抽取多个小单元比抽取少数大单元好。

由于总体的不均匀性,为了增进估计的精度,Neyman 对固定样本大小导出了最优分配的分层抽样。最优分配的推导有两层重要意义,一是分层抽样不限于传统的按比例分配,二是随机抽样不限于等概率抽取,后者对尔后的抽样设计有着深远的影响。

Neyman 在调查抽样中所作的随机化努力,可与 R. A. Fisher 在实验设计中关于试验单元安排随机化的倡议相比美。

抽样方法能否为人们所接受的一个根本性问题,是要看能否对抽样估计的精度或可信度给以科学的描述。从 Laplace 到 Bowley 都曾借助于正态分布律去计算抽样误差,却未能对估计值的精度作出有效的解释,直至 Neyman(1934)才成功地建立至今为人们所奉守的置信区间理论,置信区间已经成为一般统计学的基本概念,但它却是在研究抽样的代表性这个历史性问题中产生的。

2.4 美国普查局的工作

Neyman 以后的发展,要集中到一二个概念上来讨论,或围绕着少数作品上来分析,就比较困难了。第二次世界大战前夕,关于抽样理论及其在社会经济研究方面的应用,发展中心已由欧洲转移到美国,尤其集中反映在美国商业部普查局所做的工作中。这个时期直至20世纪60年代的主要发展大致有以下几个方面:

多级抽样设计及优化问题,包括不同单元大小和不等抽取概率的决定和实现;

系统抽样的理论和经验研究;

各种非抽样误差的经验研究;

各种复杂设计所涉及的成本函数研究;

研究时间序列的轮换设计;等等。

值得注意的,在这个时期里,由 P. C. Maha Janobis 领导的印度统计学院做了许多和美国普查局相平行的工作。

一个抽样设计要付诸实施,往往需要耗费巨额的资金和人力,没有政府和大企业的支持是难以办到的。这就是为什么抽样设计的发展,随着它的日益复杂化和大型化,要以一个行政机构或某些项目作为其代表的原因。另一方面,费用的考虑越来越突出,随着抽样设计的复杂化,往往不是在给定总样本大小下求(无偏)估计值有最小方差的最优分配计划,而是在给定费用的条件下求最小方差设计,以致一般求损失最小的抽样设计。Hansen(1909)把抽样调查看作"为达到既定目的的最有效地利用有限资源的一个经济生产过程"。

2.5 抽样设计的演变

(1) 从非概率抽样到概率抽样。表面上看来是使抽样方法摆脱人们的判断,实际上仅是在一定的分层或分群的范围内当判断不再起作用或只起有害作用时,抽样才好独立于判断。

(2) 从等概率抽样到不等概率抽样。需要有足够的辅助信息,确定对什么样的单元按多大的概率去抽取,人们不禁要想到 Bowley 所描述的目的性抽样。如何能使控制量的群均值近似于总体均值? Hansen 和 Hurwitz 猜想,单元越大,Y 值就会按比例地越大,因此提出"与单元大小呈比例"的概率抽样,从而减少被抽中单元的均值(无偏估计)对总体均值的偏离,这又何尝不是一种目的性抽样呢?

不等概率抽样是一种有偏误的抽样,不仅要有抵消偏误的估计方法,而且对于不回置的不等概率抽样,如何实现的困难问题,引起了不少的研究兴趣。

(3) 从时间上的一维抽样到多维抽样(即从单个时刻到多个时刻的抽样)。多维抽样包括多相抽样,轮换抽样(还有多级抽样?)等;无应答追踪设计、滚雪球设计等都属于多维抽样;不固定样本大小的序贯设计也属于这个范畴。多维抽样是一个动态过程,既可以把总体本身看作一个动态过程,也可以把对固定总体的认识看作一个动态过程。多维抽样的设计应充分利用对总体的逐步认识以及对未来的预见。

2.6 混合、连接、不等深度抽样

未来的调查样本设计将怎样发展?看看 Stephan 所建议的三个方向是有趣的和有启发的。

(1) 随机与系统混合抽样。例如,可利用系统样本定义一系列邻域,从每一邻域随机抽取一个单元。系统样本中的单元除非再次被抽到,否则不进入最后样本。

这种抽取方法,经适当修改可适应于分层、多级抽样。例如,既可排除相邻单元的抽取,也可按某种规则减少相邻单元的抽取概率。目的是要在充分利用已有的总体结构知识的限度内减少误差。

(2) 连接(nexus)抽样。总体中的元素除了有一种"物以类聚"的现象外,还有一种内在联系。例如研究人类关系、亲缘关系、交通运输系统、神经系统、气象站系统、结晶学等,就要考虑这种内在联系。可用图论和网络来描述并分析这种内在的相互关系。用点(顶点)表示系统的元素,用连接点的弧或有向线段表示关系。弧附有各种属性如能量,成本等。……向某一结点收集信息,不仅是为了这一个结点,也可能是为了别的结点。滚雪球抽样(Goodman,1961)就是连接抽样的一例,它从一个结点的调查开始,逐步转到同它有较密切关系的那些结点上。在连接抽样的设想中,用得着不等概率抽样,分群抽样和控制抽样。连接抽样的理论基础尚待建立。

(3) 不等深度(graduated)抽样,用于对总体中元素的变化过程的调查研究,如慢性病、农作物生长、生态学等研究。复杂性在于每个元素的变化率不是均匀的,起点也不一样。从一个变化过程来说,资料往往是过去的,对现在来说总是不完全的。对变化过程抽样的一个基本的理论和实际问题,是选择一个横截面进行调查呢?还是对一些元素作连续或重复观测呢?后者还涉及样本的损耗将随时间的迁移而递增的问题。因此,需要把现在的抽样概念加以扩充,包括规定对总体中每一单元的调查深度。比方说,对某些单元调查可限于抽样框所提供的信息,对另一些单元则要求取得一组标准的数据,而再对另一些单元则要求针对调查的焦点问题,取得一整套自始至终的发展过程的信息。亦即对总体中的一切单元获得不同深度的数据资料。

3. 估计理论

3.1 抽样的准确性

20 世纪初,Kiaer 就意识到抽样方法的推行,必须对抽样误差有所说明。但他想到的办法是同时进行两个或多个抽样调查以资比较,或将抽样结果和已有的普查资料作一对比。因此这个问题得不到实质上的解决。

Bowley 强调概率论在抽样中的作用,考虑了中心极限定理在设计总体均值或总体比率时的应用,例如他说(1926):"当我们说最大似然的平均工资估计是 24 \$ 时,偏离这个值 4 d 以上和以下的情形是等可能的,因为标准差是 6 d;超过 24 \$ · 8 d 的可能性是 1/10;超过 26 \$ 的可能性是 1/100000。如果把标准差当作一种准确度的衡量,这些结果可简单地写成 24 \$ ±6 d。"Bowley 不仅对总体均值和方差感兴趣,他还利用逆概率方法导出总体特性的后验分布。

直至 Neyman 之前,抽样调查的估计理论主要是建立在 Bayes 逆概率定理基础上的点估计。Neyman 摆脱 Bayes 定理,创造了区间估计的理论,他考虑总体特性 θ 的最小方差线性无偏估计 θ' 的渐近正态性,并利用服从 t 分布的统计量

$$t = \frac{\theta' - \theta}{S_{\theta'}}$$

(其中 $S_{\theta'}$ 代表 θ' 的标准误差)给出对应于置信度 ε 的置信区间:

$$\theta' - S_{\theta'} t_\varepsilon \leqslant \theta \leqslant \theta' + S_{\theta'} t_\varepsilon$$

这个区间不依赖于 θ 的先验分布知识。鉴于只需要有一个样本就足以说明 θ' 的可信度,于是 Neyman(1934)写道:"如果我们对一个总体特性感兴趣,并且所用的抽样和估计方法容许我们对每一可能的样本 $\sum$,定出一个置信区间 $X_1(\sum), X_2(\sum)$,不管总的未知性质是什么,都能使得下述不等式

$$X_1(\sum) \leqslant X \leqslant X_2(\sum)$$

的错误频率不超过预定的界限 $1-\varepsilon$;我们将把这样的抽样方法叫做代表性抽样,并把这样的估计方法叫做一致性估计。"

Neyman 的区间估计理论在消除人们对使用抽样方法的疑虑方面,起

到了十分积极的作用。

3.2 中心极限定理应用质疑

人们认为$(\theta'-\theta)/S_\theta$渐近于正态分布,是因为有中心极限定理作为后盾。但中心极限定理只对不断增大的总体和不断增大的样本有效。对特定大小的样本和总体,中心极限定理并没有说一个样本均值(比方说)是怎样分布的,也没有说一个估计值的误差和它的标准差的联合分布是什么。Hajek 的极限定理并不解决这个具体问题。O. Kemovhorne 举出了极端的例子,说明区间

$$\overline{X}-3S_{\overline{X}}, \quad \overline{X}+3S_{\overline{X}}$$

包含总体均值的频率可能等于零。

如果是按不等概率抽样,则问题更为困难。更不用说,某些方差估计(如 Horvitz 和 Thompson)可能取负值,将怎样构造置信区间呢?

统计推断通常是从一个纯数学的分布中抽取随机样本出发的。所谓随机样本,是指能够无限多次重复地得到和它相类似的一个样本,但事实上,我们是从一个有限总体中作往往是不回置的随机抽样。把有限总体看作来自一个超总体(Superpopulation)的样本,是企图从概念上克服这个困难的一种尝试。

3.3 抽样调查的基础研究

Fisher 的参数估计理论是建立在总体服从于某个分布的假设基础上的。当总体是有限总体时,Fisher 所引进的一些概念如似然性、充分性、信息量等是否有重新定义的必要?其意义又如何?这些问题从 50 年代末开始,已受到较多的注意。一些概念对有限总体来说是困难的,其意义如何也有待研究。例如关于有限总体的似然函数的两个极端相异的定义是:

(1) 总体中的 N 个元素可划分为少数的 k 类,每类各占比率 $p_1, p_2, \cdots, p_k, \sum p_i = 1$。抽样结果有 n_i 个元素属于第 i 类,$i=1,2,\cdots,k$,则似然函数为(令 $\sum n_i = n$)

$$\prod_{i=1}^{k} C_{Np_i}^{n} / C_N^n$$

这个函数是能够对总体提供一定信息的,例如可以按概率的大小对各种各样的总体结构作一拟合优度的排队。

(2) 假定总体 x 值为 $x_1, x_2, \cdots, x_N$。在给定样本 $x_{a_1}, x_{a_2}, \cdots, x_{a_n}$ 的条件下,总体值 $x_1, x_2, \cdots, x_N$ 的似然函数

$$L = \begin{cases} 1/C_N^n & \text{如果总体的 } x \text{ 值含有样本的 } X_a \text{ 值} \\ O & \text{如果不然} \end{cases}$$

这样的似然函数将不会对未观测到的 x 值提供任何信息(也有人持不同看法)。

这里引起一个难以解释的疑问。对于测量值 x,可以认为 $x_1,\cdots,x_N$ 是全异的。这时,如果把总体划分为少数的几组如:$X \leqslant X_k, X > X_k$;或 $X \leqslant X_h, X_h < x \leqslant X_k, X_k < X$;等等,似然函数还能对总体的未观测部分提供信息。但如果把分组数继续增大,直至每组最多含有一个总体 X 值时,似然函数就要变成无信息的了。经典的参数估计理论对有限总体说有没有意义?到底什么是有限总体的参数?把总体的 $X_1, X_2, \cdots, X_N$ 看作参数意义何在?

有限总体的元素是可以识别的。因此,在回置抽样中,当我们抽到两个相同的数值时,我们能够知道这是同一个元素两次被抽到,还是抽到有同一数值的两个不同的元素。Basu(1958)证明在回置随机抽样中,作为总体均值估计,取样本中的不同元素的平均,比取整个样本的平均更为有效(有较小方差)。按照 Basu 所下的定义,不同元素的数值(同一单元被重复抽取时只算抽取一次)构成了充分统计量。

4. 结束语:理论与实践

Neyman 在 1934 年发表的分层抽样最优分配公式,早在 1923 年已由俄国统计学家导出。由于当时俄国与西欧的隔阂,未为外界所知。有人认为,20 世纪 20 年代以前,调查统计的发展中心在俄国。挪威统计学家 Kiaer 所倡导的代表性调查,早在 19 世纪下半叶俄国就有不少的实践经验。19 世纪末 20 世纪初,俄国数理统计学家关于抽样调查的论述均属概率抽样。因此,认为当时俄国在调查统计方面的成就居世界之首,不是没有道理的。只因尔后苏联当局拒绝概率论在社会经济统计方面的应用,抽样调查的理论研究才陷于停顿。

20 世纪 40 年代以后,美国由于商业、行政管理与国防的种种需要,朝野一致,大兴调查之风,不惜耗费巨额资金。较优越的实践条件使美国代替了欧洲,成为抽样调查理论和方法的发展中心。

抽样调查的实践和理论发展是密切相关的。我国当前四个现代化需要于调查统计者甚多,有待于抽样调查的项目何止千万,这是发展理论和方法研究的极好时机。我国数理统计工作者应予高度珍视！一方面，应采用最先进的抽样调查方法以改变实践落后于理论的局面，另一方面，结合我国的实践必然产生有待研究的新理论和方法问题。至于抽样调查的基础理论，尚存在很大缺陷，有待全世界的统计学家去努力弥补。

怎样在加速寿命试验中构造置信区间[①]

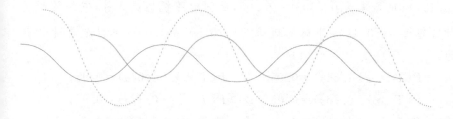

1. 问题：通过寿命试验确定一批元件（如灯泡、电子管、电机、滚珠等）的平均寿命是否达到 μ_0 小时。

假定元件的耐用时数服从正态分布 $N(\mu,\sigma^2)$，其中 μ 未知，而 σ^2 已知。设 $x_1,x_2,\cdots,x_n$ 是这个分布的一个容量为 n 的随机样本，样本均值为 $\bar{x}=\sum x_i/n$。由于 $\bar{x}$ 服从正态分布 $N(\mu,\sigma^2/n)$，故有 95% 的把握说，总体均值 μ 落在 $(\bar{x}\pm1.96\sigma/\sqrt{n})$ 之间。或写为

$$P(\bar{x}-1.96\sigma/\sqrt{n}<\mu<\bar{x}+1.96\sigma/\sqrt{n})=0.95$$

这个有 0.95 把握或置信系数的区间对称于 $\bar{x}$。但若我们的兴趣在于确定 μ 不小于某个值，则可采取单侧置信区间；以 95% 的把握说，μ 大于（等于）$\bar{x}-1.64\sigma/\sqrt{n}$。或写为

$$P(\bar{x}-1.64\sigma/\sqrt{n}<\mu<+\infty)=0.95$$

这样计算的单侧置信区间虽然能给 μ 以较高的一个下限，但并没有能够很好地解决寿命试验的问题。一则，不管时间延续多么长，都要做完样本中每一元件的寿命试验，直至它失效（寿命终止）为止，故时间上很不经济。二则，不管样本容量多么大，平均寿命的置信区间的下限总比 $\bar{x}$ 小；只要 $\bar{x}$ 一样，对相差很大和相差很小的 n 个样品寿命是不加区别的，以致损失一些有用的信息。（n 较小时也不便计算样本方差 $s^2=\sum(x_i-\bar{x})^2/(n-1)$ 以得到 t 分布）

2. 为了节约试验时间，试验只进行到一定时间为止，我们需要用另一种方法确定平均耐用时数的下限。我们知道，按照正态分布，随机抽试一

[①] 原文发表于《统计与管理》1980 年第一卷第一期。

件，它的耐用时数超过 $\mu+1.64\sigma$ 小时的概率为 5%；超过 $\mu+0.76\sigma$ 小时的概率为 $0.2236=\sqrt{0.05}$，故同时随机抽试二件，它们的耐用时数都超过 $\mu+0.76\sigma$ 小时的概率为 $(\sqrt{0.05})^2=0.05$；同理，抽试一件耐用时数超过 $\mu+(-0.124)\sigma$ 的概率为 $0.5493=\sqrt[5]{0.05}$，故同时随机抽试 5 件，它们的耐用时数都超过 $\mu-0.124\sigma$ 小时的概率为 $(\sqrt[5]{0.5})^5=0.05$。可以把上述计算写为：

$$P(T_1>\mu+1.64\sigma)=0.05 \quad 或 \quad P(T_1<\mu+1.64\sigma)=0.95;$$
$$P(T_2>\mu+0.76\sigma)=0.05 \quad 或 \quad P(T_2<\mu+0.76\sigma)=0.95;$$
$$P(T_5>\mu-0.12\sigma)=0.05 \quad 或 \quad P(T_5<\mu-0.12\sigma)=0.95。$$

其中 T_k 表示同时试验 k 件皆不失效的试验经历时数。由此得到相应的置信区间：

$$P(\mu>T_1-1.64\sigma)=0.95 \quad 即\ 0.95\ 置信区间为\ \mu>T_1-1.64\sigma;$$
$$P(\mu>T_2-0.76\sigma)=0.95 \quad 即\ 0.95\ 置信区间为\ \mu>T_2-0.76\sigma;$$
$$P(\mu>T_5+0.12\sigma)=0.95 \quad 即\ 0.95\ 置信区间为\ \mu>T_5+0.12\sigma。$$

可以利用这些置信区间作加速寿命试验。设所希望达到的平均耐用时数 $\mu=1000, \sigma^2=100^2$。为了使 $\mu>1000$ 有 0.95 的置信系数，可取

$$T_1-1.64\sigma=T_1-164=1000 \quad\quad 即\ T_1=1164;$$
$$T_2-76=1000 \quad\quad 即\ T_2=1076;$$
$$T_5+12=1000 \quad\quad 即\ T_5=988。$$

就是说，为了检验平均寿命是否达到 1000 小时，可以采取不同的试验方案。如果只用一个元件做试验，则需要做 1164 小时且不失效的试验；如果同时用 2 件做试验，则需要做 1076 小时的试验而无失效者；如果同时用 5 件做试验，则只需要做 988（小于 1000）小时的试验而无失效者。这样就有 95% 的把握说平均寿命不小于 1000 小时。这里我们看到了如何用数量换取时间以达到加速完成寿命试验的目的。当然，只在时间比元件数量相对地说更为宝贵的情况下，我们才会这样做。

显而易见，适当调整我们的计算，选用对应于不同概率水平的 σ 倍数，就能达到我们需要的任何置信水平，并不限于 95%。

虽然 σ^2 不能准确地知道，但一般可根据变异系数 σ/μ 的变化范围，对 σ^2 作一个上限估计，其中 μ 可取为希望的平均耐用时数。

3. 当然，同时用 k 件产品进行一定时间的寿命试验，不一定全无失效者。可能其中 r 件在此时间前失效（寿命终止）。这时可根据二项分布算

出 k 件中有少于等于 r 件失效的概率,从而得到相应的置信系数。

元件在时刻 T_0 前失效的概率为

$$p = \frac{1}{\sqrt{2\pi}} \int_{-\infty}^{z_0} e^{-z^2/2} dz \quad \text{其中} \quad z_0 = \frac{T_0 - \mu}{\sigma}$$

在时刻 T_0 前 k 件中有少于等于 r 件失效的概率为

$$P(y \leqslant r) = \sum_{y=0}^{r} C_k^y p^y (1-p)^{n-y}$$

相应的置信系数为

$$1 - P(y \leqslant r)$$

例:同时用 5 个元件进行 988 小时的寿命试验,发现其中 1 件在 988 小时以前失效,求 $\mu > 1000$ 的置信系数。

解:先作变换

$$z_0 = \frac{T_0 - \mu}{\sigma} = \frac{988 - 1000}{100} = -0.12$$

元件在 988 小时前失效的概率为

$$p = \frac{1}{\sqrt{2\pi}} \int_{-\infty}^{-0.12} e^{-z^2/2} dz = 0.45$$

故

$$\begin{aligned}
P(y \leqslant 1) &= \sum_{y=0}^{1} C_5^y p^y (1-p)^{5-y} \\
&= C_5^0 (0.45)^0 (0.55)^5 + C' (0.45)(0.55)^4 \\
&= 0.05 + 0.20 = 0.25
\end{aligned}$$

所求置信系数为

$$1 - 0.25 = 0.75$$

以上计算结果或表示为:

$$P(T_5^1 > \mu - 0.12\sigma) = 0.25;$$
$$P(\mu > T_5^1 + 0.12\sigma) = 0.75;$$
$$P_f(\mu > 988 + 12 = 1000) = 0.75。$$

其中 T_5^1 表示同时抽试 5 件有 1 件失效的试验经历时数。P_f 表示置信系数或置信概率。

注意,在加速寿命试验中,试验时间的长短要预先规定下来。只有这样,计算出来的置信系数才是有效的。不能根据试验结果随意改变试验进行的时间;既不能因为在预定时间内未发现有失效而延长试验时间,也不能因为在预定时间内发现有失效者而把时间缩短到未出现失效之前,作为无失效情形计算。任何事后更改试验时间的计算方法,都将是无效的。

当然,我们可以规定不止一个时间来观测失效个数,于是变成一种序贯方案,计算方法将有所不同。

以上计算置信区间的原理,不限于正态分布的总体假设,还可以考虑指数分布和韦伯分布等。

参考资料:林少宫编,农业机械试验设计与统计方法交流会补充讲义,1979年9月,湖北省机械研究所印。

附件：

1988年3月31日全国工科院校应用概率统计委员会常委扩大会议全体与会人员给林少宫教授的致敬慰问信

敬爱的林少宫教授：

  我们委员会的常委扩大会议已于三月卅一日在京圆满闭幕了，余明书同志转达了您对大会的关怀，我们全体代表向您表示感谢。

  听说您近日身体欠佳，我们都非常关切，向您致最亲切的问候！祝您早日恢复健康。

  我们委员会从筹备创建到如今，只有五年时间。但这段时间内，我们的工作有了很大的发展，各方面工作都有了不少成绩，这些都是在您的领导与辛勤操劳下取得的，我们永远不会忘记。现在您是我们的名誉主任，同志们都相信您会一如既往，像过去一样领导着我们前进。

  目前形势很好，许多工作等待我们去做，其中最迫切的两件工作是承接课题开展科技协作和开展国际学术交流活动，由于您在这两方面有着很高的声望，将起着别的任何人所不可能起的巨大作用，我们热切期望并坚决相信在您的领导与关怀下，我们应用概率统计委员会的工作会获得更大成功！

  敬您身体健康，万事如意！

候振挺（签名）
陶宗英（签名）
刘文（签名）
沈恒范（签名）
韦博成（签名）
马逢时（签名）
赵维谦（签名）
余明书（签名）
赵达纲（签名）
杨振海（签名）
郭福星（签名）

蒋承仪(签名)
关颖男(签名)
吴让泉(签名)
胡乃冈(签名)
张魁元(签名)
盛承懋(签名)

一九八八年三月卅一日于北京

第三部分

正交试验设计

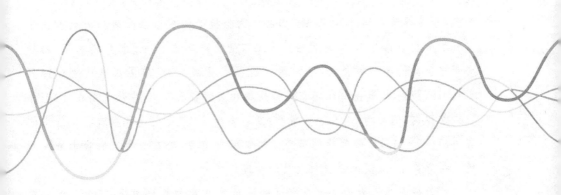

"实(试)验设计是数理统计的一个专门领域,析因设计及其部分实施又是实验设计中的极具应用性而比较抽象的一类理论成果(但鲜为人知,中外皆然)"

——林少宫

"林老对正交试验设计方面的研究,成果丰富而又集中,多属国内首创,并独具中国特色"

——《奋发》[①]

◇ ◇ ◇

1972年,林少宫教授在工业上推广应用正交试验设计,从因数筛选到定向探索,再到优化推进,拟定了一套简单易懂而又有普遍应用意义的模式,在武汉化工研究所取得优异成绩。1974年暑期,林教授在武汉市科委为工矿企业举办的"正交试验设计应用讲习班"担任主讲。1975年春,林教授一行考虑在广东在农业上应用正交试验,考察了花县、番禺、肇庆、新会、台山、信宜等地。之后,把正交试验从工业推广到农业的普及应用上,在湖北潜江、沙市、荆州、沔阳、宜昌、孝感等地,在"农村和田野"现场,进行了卓有成效的普及和推广,在多个项目中取得成果。1976年,林教授根据实际应用经验,利用极差代替F检验,与吕梓琴合作,成功编制了"正交试验极差临界值系数表",大大加速了实验的分析与计算工作,在我国广大工农业生产实验中一度被广泛采用,给社会带来巨大的经济效益,后来被第三机械工业部编入《应用数学成果巡回展览》。1977年,林教授前往广东省科技局和湖南省农科院(袁隆平所在处)等单位讲学咨询,随后为《湖北农业科学》撰写系列论文"正交设计在农业试验上的应用(1—9讲)",诠释部分析因设计的原理与应用,给农业生产试验带来了积极的指导作用,深受基层农科人员的欢迎。

林教授撰写了大量关于正交试验设计的文章和普及读物,其中一部分被本章收录。

[①] 《奋发》的副标题是"华中科技大学经济学院发展历程",此书于2014年由华中科技大学经济学院编印。

正交试验设计讲座[①]（农机试验例解）

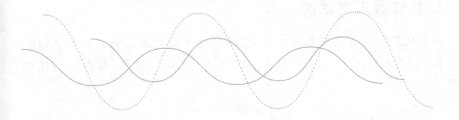

1. 什么是正交试验设计

正交试验设计是安排多因素试验的一种方案。多因素试验由于因素之间有交互作用而有一定的复杂性，往往进行多次试验而分不清楚每个因素的作用大小及其主次关系，选不出较好的配方或工艺条件来。正交设计的特点在于：

（1）试验次数少；

（2）因素作用的分析计算简单，因而能较快地找出好的配方或工艺条件。

为什么叫正交设计？正交就是垂直的意思。打个比喻。直角三角形的斜边可以分解为互相垂直的两个对边。把斜边比作试验的总效果，互相垂直的两边比作两个不同因素的分效果。我们知道，斜边边长的平方等于互相垂直的两个边长的平方和。这样，总效果就清清楚楚、不折不扣地分解为两个不同因素的分效果之和。正交设计就起到这种正交分解或者说垂直分解的作用。这一分解在多因素试验中大有用处（实例略）。

正交设计的推广应用是20世纪70年代末的新生事物，由于社会主义制度的优越性，它在我国普及范围之广，见效之快，世界各国没有一个比得上。

正交设计的推广应用，已在全国范围内获得巨大的成功。这是我们实行了三结合的结果。第一，组织领导重视，有组织有计划地抓推广应用工作。第二，走群众路线，紧密结合群众的生产经验和专业知识。第三，数学

[①] 原文发表于《湖北机械》1976年第4期。

工作者走与工农相结合的道路,数学方法一再简化、表格化,使正交试验设计做到"简单、有效",适应广大工农群众的需要。任何试验项目,要应用正交设计取得成功,都要贯彻执行三结合的方针。

2. 怎样做正交试验

正交试验是通过正交表安排试验,所以也叫正交表试验。正交表如常见的表 $L_8(2^7)$、$L_{16}(2^{15})$、$L_9(3^4)$、$L_{27}(3^{13})$、$L_8(4\times2^4)$ 等等,都由一个模样的符号

$$L_N(S^K)$$

来表示。这里 L 表示正交表,N 表示试验次数(处理组合数),K 表示最多能安排的因素个数,S 表示因素的水平数。现在通过一个实例看怎样利用一张正交表来安排试验。

例1:水田收获机械行走机构性能试验(取自一机部农机研究所《农机情报资料》1975 年 1 月)。

试验目的:合理选择整机参数,提高行走机构通过性能。

用三个指标评定性能好坏:

(1) 行走阻力,越小越好;

(2) 滑转率,越小越好;

(3) 下陷深度,越浅越好。

影响指标的三个主要因素被认为是:①接地压力;②履带板型式;③重心位置。(其他因素如行驶速度暂固定在 2 公里/小时)。

进行哪些试验呢?也就是,取多大的接地压力,用什么型式的履带板,以及选怎样的重心位置来做试验呢?正交试验要求我们对每个因素各选定若干个数量或状态,用试验设计的术语说,各取几个水平,以便安排试验。本例中,每个因素各取三个水平,从而制定一张因素水平表,如表1(暂不看区组一列)。这样就可按各种水平搭配(或叫处理组合)来进行试验。

表 1 水田收获机械行走机构性能试验

水平	A 接地压力	B 履带板型式	C 重心位置	D 区组
1	0.18 kg/cm²	间隔:无	中点前(120 mm)	东段
2	0.21 kg/cm²	间隔:大	中点	中段
2	0.23 kg/cm²	间隔:小	中点后(120 mm)	西段

比如说,压力取 0.21,间隔取"无",重心取"中点";压力取 0.23,间隔

取"无",重心取"中点前";等等。可能的水平搭配很多,共 $3\times3\times3=27$ 种,一一进行试验,往往条件不允许。正交表告诉我们选做哪些试验,不一定要全做。如果将因素排好队,水平编好号,像表1那样,就可以对照正交上的号码安排试验。比如,按表1,"压力取0.18,间隔取'大',重心取'中点'"可读作"1,2,2"。反过来"2,2,3"则表示"压力取0.21,间隔取'大',重心取'中点后'"。

现把压力、履带板、重心三个因素依次排在正交表 $L_9(3^4)$ 的前三列,如表2所示,这三列的每行数码各表示什么,可想而知。

还有一个问题,试验要在田地里做,有些田地的土质松软,有些则坚实,为了使试验结果的分析比较不受土质差异的影响,最好把试验田地适当地划分区组:东、中、西三段,每段算作一个区组,利用正交表 $L_9(3^4)$ 上的空列(第四列)把区组当因素一般排上去,区组列的三个水平分别指东段、中段和西段(见表1)。每次试验怎样进行,通过正交表的安排都清楚了。例如表2第3号试验是:

"3,1,3,3"表示"压力取0.23,间隔取'无',重心取在中点后120 mm,在田地的西段进行试验。"表中有9个试验号,一共进行9次试验。

表2 正交表 $L_9(3^4)$ 试验

试验号	比压 A	履带 B	重心 C	区组 D	指标			加权/%			综合评分 $\bar{y}$
					阻力/千克 y_1	滑转率/% y_2	下陷/毫米 y_3	50% y_1-615	30% $y_2-2.3$	20% y_3-8	
1	1(0.18)	1(无)	1(前)	1(东)	638	4.1	8.0	23	1.8	0	12.64
2	2(0.21)	1	2(中)	2(中)	632	3.3	10.7	17	1.0	2.7	14.94
3	3(0.23)	1	3(后)	3(西)	816	9.1	10.6	201	6.8	2.6	79.54
4	1	2(大)	2	3	681	5.5	10.3	3	3.2	2.3	33.90
5	2	2	3	1	838	9.4	15.5	223	7.1	7.5	98.30
6	3	2	1	2	773	6.5	14.0	158	4.2	6.0	67.90
7	1	3(小)	3	2	627	2.3	10.6	12	0	2.6	9.30
8	2	3	1	3	615	4.4	11.8	0	2.1	3.8	18.60
9	3	3	2	1	632	5.8	12.5	17	3.5	4.5	30.10

续表

试验号	比压 A	履带 B	重心 C	区组 D	指标 阻力/千克 y_1	指标 滑转率/% y_2	指标 下陷/毫米 y_3	加权/% 50% y_1-615	加权/% 30% $y_2-2.3$	加权/% 20% y_3-8	综合评分 $\bar{y}$
阻力 K_1	1946	2086	2026	2108							
阻力 K_2	2085	2292	1945	2032							
阻力 K_3	2221	1874	2281	2112							
阻力 k_1	648.7	695.3	675.3	702.7							
阻力 k_2	695.0	764.0	648.3	677.3							
阻力 k_3	740.3	624.7	760.3	704.3							
阻力 R	91.6	139.4	112.0	26.7							
阻力 优水平	A_1	B_3	C_2								
滑转 k_1	4.0	5.5	5.0	6.4							
滑转 k_2	5.7	7.1	4.9	4.0							
滑转 k_3	7.1	4.2	6.9	6.3							
滑转 R	3.1	2.9	2.0	2.4							
滑转 优水平	A_1	B_3	C_2								
下陷 k_1	9.6	9.8	11.3	12.0							
下陷 k_2	12.7	13.3	11.2	11.8							
下陷 k_3	12.3	11.6	12.2	10.9							
下陷 R	3.1	3.5	1.0	1.1							
下陷 优水平	A_1	B_1	C_2								
综合评分 k_1	18.6	35.6	33.0	47.0							
综合评分 k_2	43.9	66.7	26.3	30.7							
综合评分 k_3	59.1	19.3	62.3	44.0							
综合评分 R	40.5	47.4	36.0	16.5							
综合评分 优水平	A_1	B_3	C_2								

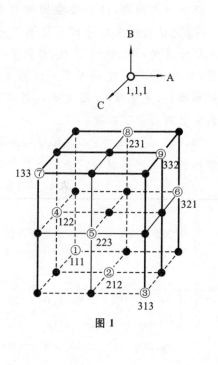

图1

试验做完后,有了全部指标值,就要进行分析计算。虽然本例是多指标情形(三个指标),但分析正交表试验的基本方法,是对一个指标的分析方法。假定只有阻力一项指标,我们将做如下的分析。

直接比较9次试验结果,阻力越小越好,那么8号试验最好,7号也差不多好,能不能就此下结论说,8号或7号提供了最好的整机参数呢?有问题。一则试验有误差(包括土质差异),二则8号或7号还不能代表27种可能水平搭配中最好的。因此还要进行统计分析。

统计分析:逐列按不同水平计算试验结果的平均值k及其极差R(见表2)。例如第1列的1水平平均值计算如下:

$$K_1 = 638 + 681 + 627 = 1946, \quad k_1 = \frac{K_1}{3} = \frac{1946}{3} = 648.7$$

第1列的水平平均值的极差是指该列的最大水平平均值-最小水平平均值$= 740.3 - 648.7 = 91.6$,这极差代表该列因素(比压)的水平变化(从0.18变到0.23)所起的作用(或叫效应)。

首先,按各列水平平均值的极差大小判断因素作用的主次顺序为:B(履带板) > A(比压) > C(重心),并由第4列极差知区组差异比较小。

其次,对较主要的因素,按其水平平均值的大小选较优水平,平均值越小越好,故依次选为B_3、A_1、C_2,即履带板间隔以3水平(小间隔)好;比压以1水平(0.18 kg/cm^2)好;重心以2水平(在中点)好。这一组水平搭配究竟和8号或7号不一样,也不在表$L_9(3^4)$的9次试验之列。

以上分析,建立在逐个因素优选的基础上,在一定程度上忽略了因素间的交互作用。但通过尔后的验证比较试验,可以证实分析结果是否有效。由于参考了8号和7号这两号的较好试验条件,交互作用也没有完全被忽略。

3. 正交设计的原理

3.1 均匀分散性

用立体直角坐标代表三个因素,取原点为水平组合(1,1,1),27个水平组合有如立方体中的27个点(实圆点加空圆点)(见图1)。正交表$L_9(3^4)$从中选取9个点(空圆点)。这9个点分布得很均匀,纵、横、直向每个截面上都有3点,相对于27个点来说很有代表性;27个点中有好有坏,那么,这9个点中也会有好有坏。优选法的目的,就在于试验点数不多而能碰上好点。

3.2 整齐可比性

为什么在履带板和重心同时变动的情况下,能单独比较比压的三个水平平均值,从而估计比压的作用并优选比压的水平呢?因为比压的1水平同履带板的三个水平各搭配一次;比压的 2 水平同履带板的三个水平各搭配一次;比压的 3 水平也是如此。同样,比压的 1 水平同重心的三个水平各搭配一次;比压的 2 水平同重心的三个水平各搭配一次;比压的 3 水平也是如此。比压的每个水平同区组的各个水平的搭配也是如此整齐的。因此,比压的三个水平平均值的差异主要反映了比压的三个水平的好坏。就是说这三个水平平均值是可以比较的。

水平的这种搭配关系对比压这个因素是如此,对其他因素也是如此。

3.3 潜在重复性

从表 2 看,我们对每个因素的每个水平都做了三次试验,三次试验的平均值比单独一次试验值更能反映某一水平的好坏。统计分析通过平均值的计算而增加了分析结果的准确性。

单从提高准确性看,在没有交互作用的情况下,正交表 $L_9(3^4)$ 试验,对每个因素的每个水平都不折不扣地起到三次重复试验的作用。对于三个因素各三个水平,就起到了 3×3 的三次重复即 27 次试验的作用。

4. 多指标问题

本例是多指标情形,处理方法有二:

(1) 逐个指标进行分析,然后加以综合评定;

(2) 先对全部指标做综合评分,再对评分结果进行如同单指标那样的分析。

一方法是逐个指标分析法。前面已分析了阻力,同样方法(见表 2),对滑转率,根据 k 和 R 值的计算结果,按因素的主次顺序,其较好条件为 $A_1B_3C_2$;对下陷深度则为 $B_1A_1C_2$。

阻力、滑转率、下陷三个指标的分析结果矛盾不大,只有对阻力和滑转率宜取 B_3 而对下陷宜取 B_1 之差。平衡一下利弊,较好条件可定为 $A_1B_3C_2$,即仍然是比压选取 0.18;履带板间隔宜"小";重心放在中点好。

另一方法是综合评分法。按指标的重要性,给阻力的变化幅度 838－615＝223 千克以 50 分;给滑转率的变化幅度 9.4%－2.3%＝7.1% 以 30 分;给下陷深度的变化幅度 15.5－8.0＝7.5 毫米以 20 分。也就是,从试验的

最低阻力算起,每增加 1 千克阻力给以 $\dfrac{50}{223}$＝0.22 分;

最低滑转率算起,每增加 1% 滑转率给以 $\frac{30}{7.1}=4.2$ 分;

最低下陷深度算起,每增加 1 毫米深度给以 $\frac{20}{7.5}=2.6$ 分。

这样,将每一指标值减去相应的最低指标值,就容易算出每一试验结果的综合评分(见表 2 右边四列)。例如第 1 号试验给以

$0.22\times 23+0.42\times 1.8+2.6\times 0=12.6$ 分,余类推。

综合评分越低越好。对综合评分的分析计算结果(见表 2 最底下一行)和逐个指标分析结论一致。

验证试验。从正交表试验分析出来的好条件,还要结合生产经验和专业知识来考虑,然后选出合乎实际的好条件,加以验证。

比压越低越好,是合乎推理的。但要做到比压 0.18,在机具设计上,还要付出很大努力。另外,从 9 次试验的分析结果看,虽然重心放在中点较好,放在中点前 120 毫米也不算差。同时根据实际经验,认为放在中点前 120 毫米可能较好,故再安排二个验证比较试验:

试验号　试验条件　综合评分
10　　　$A_2B_3C_1$　　78.10
11　　　$A_1B_3C_1$　　46.96

主要是为了比较 10 号和前面的 7 号试验。虽然 10 号和 11 号试验在另一块田地上进行,但 10 号和前面的 8 号条件相同,11 号和 7 号仍可通过它们间接进行比较。为了把 10 号试验的评分 78.1 分和 8 号试验的评分 50.2 分等同起来而比较 11 号和 7 号:

对 7 号试验计算　$40.9/50.2=81.47\%$

对 11 号试验计算　$46.96/78.1=60.12\%$

这两个相对评分仍然是越低越好,因此可认为 11 号优于 7 号,且定为最好的整机条件。

5. 交互作用的考察

一般只需要考察二因素间的交互作用。什么是二因素间的交互作用?大致地说,就是一个因素选什么水平好,要受到另一因素取什么水平的影响。例如两个水稻品种,一个耐肥,一个不耐肥。耐肥的要求肥料多用些,不耐肥的要求肥料少用些。到底施肥多些好还是少些好,受品种种类的影响。反过来,哪一品种较为可取,要看肥料是否充足。这就是品种和肥料

二因素间有交互作用。当二因素之间有较大交互作用时，它们的优水平必须同时选择。

当因素较多（多于三个）时，为考察两两因素间交互作用的大小，往往安排二水平表 $L_{16}(2^{15})$ 试验。

例2：插秧机栽植性能试验。

规定每穴插植株数，观察漏插情形。

指标：漏插率，越小越好。

对影响指标的因素考虑如下：

A 翻土后天数：A_1（三天）或 A_2（一天）

B 水深：B_1（浅）或 B_2（深）

C 播种密度：C_1（稀）或 C_2（密）

D 作业速度：D_1（慢）或 D_2（快）

E 爪的种类：E_1（甲）或 E_2（乙）

F 抓秧的位置：F_1（低）或 F_2（高）

待考察的因素及交互作用如图2所示。圆点表示因素，两点间连线表示两因素间的交互作用。该试验要求考察 A、B、C、D、E、F 6个因素作用及 AB、AC、AE、AF、BE、BF 及 EF 7个二因素交互作用。选用的正交表至少要有 6+7=13 列。正交表 $L_{16}(2^{15})$ 有15列，可能够用。把因素 A、B、C、D 排在表 $L_{16}(2^{15})$ 的 A、B、C、D 列上，再把因素 E 和 F 分别排在 ABD 和 ACD 两列上，这样就能做到，不但每一因素的作用都有表中的一列和它对应，而且每一交互作用也都有表中的一列和它对应。这些对应列都标明在表 $L_{16}(2^{15})$ 的表头上（见表3）。（至于怎样定出这些对应列，可参考华中工学院数学教研组编《正交设计与分析》第二章第一节。余下的二列可作为误差列，这里不作详细讲解了。）

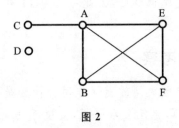

图 2

安排试验时，只管看排有因素 A、B、C、D、E、F 6列的数码。例如第5号试验是：$A_1 B_2 C_1 D_1 E_2 F_1$。该号试验结果是：漏插率＝8.5%。

表3 正交表 $L_{16}(2^{15})$ 试验

试验号	A	B	AB	C	AC	EF误(BC)	(ABC)	D	BE(AD)	AE(BD)	E(ABD)	CD	F(ACD)	误(BCD)	BF(ABCD)	漏插率/%
	1	2	3	4	5	6	7	8	9	10	11	12	13	14	15	
1	1	1	1	1	1	1	1	1	1	1	1	1	1	1	1	7.6
2	1	1	1	1	1	1	1	2	2	2	2	2	2	2	2	10.1
3	1	1	1	2	2	2	2	1	1	1	1	2	2	2	2	9.4
4	1	1	1	2	2	2	2	2	2	2	2	1	1	1	1	3.8
5	1	2	2	1	1	2	2	1	1	2	2	1	1	2	2	8.5
6	1	2	2	1	1	2	2	2	2	1	1	2	2	1	1	3.3
7	1	2	2	2	2	1	1	1	1	2	2	2	2	1	1	13.5
8	1	2	2	2	2	1	1	2	2	1	1	1	1	2	2	3.3
9	2	1	2	1	2	1	2	1	2	1	2	1	2	1	2	31.6
10	2	1	2	1	2	1	2	2	1	2	1	2	1	2	1	1.8
11	2	1	2	2	1	2	1	1	2	1	2	2	1	2	1	8.8
12	2	1	2	2	1	2	1	2	1	2	1	1	2	1	2	30.9
13	2	2	1	1	2	2	1	1	2	2	1	1	2	2	1	4.9
14	2	2	1	1	2	2	1	2	1	1	2	2	1	1	2	0
15	2	2	1	2	1	1	2	1	2	2	1	2	1	1	2	0
16	2	2	1	2	1	1	2	2	1	1	2	1	2	2	1	6.4
k_1	7.4	13.0	5.3	8.5	9.5	9.3	8.5	10.5	9.7	8.8	7.7	12.1	4.2	11.1	6.3	
k_2	10.6	4.9	12.7	9.5	8.5	8.7	8.1	7.5	8.2	8.0	10.3	5.9	13.7	6.7	10.6	
R	3.2	8.1	7.4	1	1	0.6	0.4	3.0	1.5	0.8	2.6	6.2	9.5	4.4	4.3	

对全部试验结果的分析计算方法,和例1相仿。所不同的是,在本例中,表上还标明有交互作用列。故除了计算 A、B、C、D、E、F 6个因素的作用(效应)外,还要计算待考察的7个交互作用。(本来对于6个因素可以排出 $C_6^2=(6×5)÷2=15$ 个二因素间的交互作用,但根据图2的设想,只有7个需要考察,其余的8个都可以忽略不计。)

仍然对每列算出水平平均值 k 及其极差 R,见表3。根据 R 一栏数值的大小判别各列的相对重要性。再对较重要因素比较其 k_1 和 k_2(注意表3中打圈的 k 值),从而选出较优水平。对较重要的交互作用,则选取有关因素的较优水平搭配,如表4。对较次要因素,可结合实际情况选其水平[①]。本例中,全部因素作用和二因素交互作用的主次顺序是:F、B、AB、AF、……。因漏插率越小越好,故对 F 取 F_1;对 B 取 B_2;而对 AB 则取组合 A_2B_2,……。对 AB 取 A_2B_2 的原因,是 AB 的 4 种水平搭配 A_1B_1、A_1B_2、A_2B_1、A_2B_2 中以 A_2B_2 这一搭配的平均漏插率最小,如表4所示的计算。对 AF 二因素如何搭配为好,也可作类似的分析计算。总起来,较优水平搭配是 $A_2B_2F_1$,其余因素的水平可根据实际情况选定。(注:如果不考虑交互作用,单独从 A 这列考虑,就可能选 A_1 为较优水平,那就搞错了。)

表4　AB4种水平搭配的平均漏插率　　　　　单位:%

	A_1	A_2
B_1	$\frac{7.6+10.1+9.4+3.8}{4}=7.7$	$\frac{31.6+1.8+8.8+30.9}{4}=18.3$
B_2	$\frac{8.5+3.3+13.5+3.3}{4}=7.1$	$\frac{4.9+0+0+6.4}{4}=2.8$

AB 交互作用=$[(18.3+7.1)-(7.7+2.8)]\div 2=7.4$

事实上,$A_2B_2F_1$ 这一搭配条件和14号或15号试验条件相符。本来不加分析也可以直接从16个漏插率中比较出14号或15号试验条件最好,但经过统计分析得出来的结论,意义有所不同。第一,经统计分析后,矛盾主次分明。第二,减少或排除误差的干扰(特别是经过统计检验之后,更有一定把握)。第三,14号或15号条件充其量只代表16种水平搭配中的较优,而统计分析则考虑全面试验 $2^8=256$ 种水平搭配中的较优。第四,弄清楚许多交互作用及水平搭配关系,例如,当 A 取 A_1(翻土后三天)时,水的深浅关系不大;但当 A 取 A_2(翻土后一天)时,深水就比浅水好得多。

6. 正交表的使用

正交表的编制有一定的数学原理,对每个表来说,因素个数和水平数以及可考察的交互作用,都受到一定限制。因此每选用一个表,就要受到这个表的限制。虽然如此,灵活性还是有的,因为可以在不同的表里挑选使用。通常能熟悉二水平表 $L_8(2^7)$、$L_{16}(2^{15})$,三水平表 $L_9(3^4)$、$L_{27}(3^{13})$,以及从二水平变到四水平或从四水平变回二水平(参考华中工学院《试验

---

① R 值大到多大,才宜根据 k 值的大小选优水平,可通过某种统计检验。但统计检验要对试验误差有所估计,这里不详细讲了。

设计与分析》附表)就可以了。

虽然正交表的使用普遍地节约试验次数,但每个表的节约程度不一。表上因素排得越满,节省越多,但排得越满,越难于考察交互作用。

当需要考察交互作用时,一般以二水平正交表为宜。二水平表,特别是表 $L_{16}(2^{15})$,试验次数比较节约,交互作用的分析也比较简单、明确。当然,用此表至少要有四、五个因素以上。

当水平较多或因素较多,较小的正交表不能容纳,而较大的正交表又嫌试验次数过多时,则可适当修改水平数或因素个数,或将试验分批进行,每批用一个较小的正交表。一般说来,两三个小正交表合并起来的试验次数要比一个大正交表的试验次数少。

在整个试验方案中,不要漏掉任何重要因素,水平也要选取得当。选好因素、定好水平是试验成败的关键所在。怎样选定得好,要十分重视生产经验和专业知识,要尊重群众的首创精神,搞群言堂。

不管因素是数量的或属性的,都可以用正交表安排试验。不管指标是数量的或属性的,正交试验的分析方法都是一个模样。因此,对属性指标,要将它折合数量,给以某种评分,才好进行统计分析。

指标值是分析计算的依据,正交表试验分析对这些数值反复加以利用,进行各式各样的计算和比较,可以说最充分地利用了这些数据所提供的信息了。因此,这些数据的观测,应尽可能做到准确可靠。必要时要进行重复观测或做重复试验(当然正交表试验本身有一定的潜在重复性)。有时为了得到某种指标值,还要使用一定的统计技巧。例如滑转率的田间测量数据处理,下陷深度的田间取样方法,漏插率或均匀率的合理评定等。还有多指标综合评分中的合理加权问题。总之,整个试验从安排到取数都要认真,一丝不苟。只有这样,分析结果才会有效。

正交设计法[1]

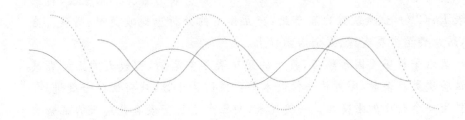

1. 概述

正交试验设计,简称正交设计,是试(实)验设计的重要组成部分。利用具有正交性的试验设计,能显著提高对试验结果的分析和计算效率。

对单因素比较试验来说,正交设计能有效地控制或排除外界因素对试验目的的干扰。对多因素析因试验来说,当某些交互作用因素可以忽略不计时,正交设计可以大幅度地减少试验次数,而不致影响对必要的因素效应的分析。即使在因素效应互相混杂的情况下,正交设计也不失为一种合理的有效的均匀布置试验点的方法。

多年来,正交试验设计作为一种多因素优选法,在国内外工、农业的许多试验中取得了显著的经济效益。

比如,有两台不同型号的机器,分别由两名工人操作,结果机器甲的加工合格率高于机器乙的加工合格率。这是由于机器的差异,还是由于工人技术水平的差异?很难说。为了判断机器甲是否优于机器乙,特别是在不十分清楚这两名工人的技术水平的情况下,最好安排他们轮流在两台机器上各操作一次或操作相同的次数。这就是一种正交设计。这样一来,简单地计算两台机器的平均加工合格率,就能够对机器作出公平的比较(当然,还可以计算两名工人的平均加工合格率以比较他们的技术操作水平)。"正交"意味着不同因素——机器与工人——之间的均衡配搭。从而使计算出来的各因素效应——机器平均加工合格率或工人平均加工合格

---

[1] 节选自《现代工程数学手册(第Ⅳ卷)》中的第六十一篇,华中工学院出版社,1987年。

率——在统计上互相独立(关于正交性与独立性见本文 3.2 节(2))。

2. 区组试验(局部控制)

费歇早在 20 世纪 20 年代研究生物与农业统计时,曾提出试验设计的三项原则:局部控制、重复与随机化。这三项原则至今为人们所奉守。所谓局部控制,就是通过试验的适当安排,控制外界因素对试验的干扰,从而减少试验的系统误差,提高试验的精度。为了达到局部控制的目的,可利用诸如随机完全区组[①]、拉丁方区组等区组设计,把较大部分的系统误差从试验结果中分离出来。

2.1 随机完全区组设计

比如,要比较 8 个小麦品种 $A, B, C, \cdots, H$ 的产量高低,决定进行 4 次重复试验。为了消除试验田的土地肥力差异对产量的影响,把试验田划分为 4 个区组,要求尽可能做到每一个区组之内的肥力均匀,以达到"局部控制"的目的。通常,要把区组内的小区安排得尽量集中,让土壤肥力差异仅存在于不同的区组之间。然后,按随机方式把每一品种安排到每个区组中的一个小区里去做一次试验,区组内的小区数等于品种数,这就是所谓随机完全区组设计,如图 1 所示。这里"完全"二字指区组内的小区数等于品种数,如果小区数少于品种数,就是不完全区组了。随机完全区组设计的基本思想,就是要做到品种与区组正交。即每一品种都在各种不同肥力条件下进行一次试验。这样一来,品种之间的产量差异就比较如实地反映了品种本身的优劣,而不能说是由土壤的肥力差异所造成。

在试验设计中,"区组"一词不限于指土地的划分,而泛指试验的外界条件如材料、人员、时间、空间或环境条件的划分。

区组 I			
A	H	G	F
E	D	C	B

区组 II			
H	F	G	B
C	A	E	D

区组 III			
B	E	H	C
F	A	D	G

区组 IV			
C	E	H	F
D	B	A	G

图 1

---

① 区组是试验设计中的一个专用名词。在试验设计中把每一个试验的操作单元叫作小区,若干试验的外界条件相近的小区组成一个区组。

2.2 拉丁方区组设计

在农田试验中,随机完全区组设计是对肥力差异的一种单向划分,它以区组形式按肥力差异划分试验地段。当肥力梯度明显呈两个方向变化时,则可采取按变化的两个方向划分成行(横)区组和列(纵)区组,称为双向区组。所谓拉丁方区组设计,就是将品种(或处理)安排在一个双向区组中,要求每个品种(或处理)在每一行区组和每一列区组中都出现一次且只有一次。因此,在拉丁方区组设计中,重复试验的次数等于品种(或处理)个数。为了说明简单起见,现在将供试验的品种减为3个,把整个试验地按肥力变化方向划分为3行和3列。为了控制行与行之间和列与列之间的肥力差异对3个品种产量的影响,要求每个品种在每一行、每一列里出现一次。满足这种要求的安排有12个之多,图2所示仅是其中之一。这种设计的基本思想,仍然是为了使品种之间的产量差异能客观地反映品种本身的优劣,而与行间、列间的肥力差异无关。因为每一品种在每行都出现一次,这就是说品种与行正交,所以行的差异消除了。同理,品种与列正交,列的差异也消除了。此外,每行和每列都相遇一次,所以行与列也是正交的。

以上设计,虽然先出现于农业试验,但容易找到工业上的应用。例如3种配方的粉末冶金压件在炉中排成拉丁方进行烧结试验。

A	B	C
B	C	A
C	A	B

图 2

2.3 正交拉丁方区组设计

假定在图2的拉丁方中,9个小区的播种由3个农民 α、β、γ 来分担,各播种3个小区。一种合理的分派方法是,不仅使每一个农民都承担每一个小麦品种的播种,而且每一个农民都能在每一行、每一列播种一次。这样就做到了:农民与品种正交;农民与行正交;农民与列正交。这种分派方法可由图3(a)来表示。图3(a)本身也是一个拉丁方。由于图3(a)中的希腊字母 α、β、γ 和图2中的拉丁字母 A、B、C 各相遇一次,这样的两个拉丁方称为一对互相正交的拉丁方。如果把它们重叠起来,就称为希腊拉丁方,如图 3(b)所示。这一设计,除了控制行、列差异外,还控制了不同农民的技术差别对产量的影响。因此,这是一种多向的区组设计。

α	γ	β
β	α	γ
γ	β	α

图 3(a)

Aα	Bγ	Cβ
Bβ	Cα	Aγ
Cγ	Aβ	Ba

图 3(b)

正交拉丁方的个数。

一个行(或列)数为 s 的拉丁方称为 s 阶拉丁方。两个同阶拉丁方,如果把它们重叠起来,则两个拉丁方中的 s 个字母联合组成 s^2 个有序对,如图 3(b)所示,若这 s^2 个有序对是不同的,则称这两个拉丁方正交。一组同阶拉丁方,若任何两个都正交,则称这组拉丁方相互正交。可以证明,s 阶拉丁方其相互正交的拉丁方数不能超过 $s-1$ 个。对于某些 s 值,比如 $s=6$ 并不存在一对正交拉丁方,但当 s 为素数或素数幂时,可用有限域方法构造出 $s-1$ 个相互正交的拉丁方。

假如把正交拉丁方行、列及每个拉丁方中的字母作为一个因素,可以组成 $m+2$ 个正交因素试验,其中 m 为相互正交拉丁方的个数。例如,3 阶拉丁方,只有两个正交拉丁方,可以组成 4 个正交因素:行、列、品种和农民。因此,构造一组个数最多的相互正交拉丁方,是组合设计中的一个课题。当 s 为非素数幂时,除 $s=6$ 外,均未得到解决。因此下面仅限于讨论 s 为素数或素数幂时,正交拉丁方和正交表的关系问题。

3. 正交拉丁方与正交表

3.1 正交表是正交拉丁方的推广

正交表是安排多因素试验的有效工具。正交表和正交拉丁方有着内在的联系。一些正交表可以直接由正交拉丁方变出。例如,对图 3(b)中的 A、B、C 三个品种依次给以代号 0、1、2(不一定是 0、1、2,其他代号比如一、0、+也行);对农民 α、β、γ 也依次给以代号 0、1、2;再用同样的代号对行和列编上号,如图 4 所示。把图中每一小区所在的行号、列号及该小区内的品种和农民代号排成一横行。小区的顺序从上到下,从左到右。第 1 行第 1 列为 0,0,0,0;第 2 行第 1 列有 1,0,1,1,…;第 3 行第 3 列有 2,2,1,0。9

个小区依次给出 9 行号码,这就构成一张 9 行 4 列的 3 水平正交表 $L_9(3^4)$①,见表 1 用虚线框住的部分,其中的不同代号将称为水平。

列:	0		1	2		
行:	品种:	农民:				
0	0	0	1	2	2	1
1	1	1	2	0	0	2
2	2	2	0	1	1	0

图 4

表 1

区号	行	列	品种	农民	产量
		L_9	(3^4)		
1	0	0	0	0	y_1
2	1	0	1	1	y_2
3	2	0	2	2	y_3
4	0	1	1	2	y_4
5	1	1	2	0	y_5
6	2	1	0	1	y_6
7	0	2	2	1	y_7
8	1	2	0	2	y_8
9	2	2	1	0	y_9

类似地,可从一个 2 行 2 列的拉丁方排出一张 4 行 3 列的 2 水平正交表 $L_4(2^8)$,如图 5 和表 2 所示,其中用"−""+"表示两个不同的水平。

列:	−	+
行:		
−	+	−
+	−	+

图 5

---

① 符号 $L_n(t^k)$ 表示一张正交表,其中 n 表示表的行数,即试验次数;k 表示表的列数,即最多能安排的试验因素数;t 为因素的水平数。

表 2

区号	行	列	品种
		$L_4(2^8)$	
1	−	−	+
2	+	−	−
3	−	+	−
4	+	+	+

一般，对于阶数为 s 的拉丁方，如果能找到它的 $s-1$ 个互相正交的拉丁方，就能将它们合成一个 s^2 行 $s+1$ 列的 s 水平正交表 $L_{s^2}(s^{s+1})$。但是并非所有正交表都能从正交拉丁方变出，如 $L_8(2^7)$，$L_{18}(2\times 3^7)$ 等都不能从正交拉丁方导出。因此，正交表是正交拉丁方的推广。

3.2 正交表的正交性

正交表的正交性表现在：任二列的每一种号码配搭都出现，并且出现同样多次。例如，表 1 的 $L_9(3^4)$ 表中，任二列都给出 $(0,0),(0,1),(0,2)$，$(1,0),(1,1),(1,2),(2,0),(2,1),(2,2)$ 这九种可能的号码配搭各一次。正是这种性质，使试验点（指因素的水平组合）在试验范围（指因素的水平变化范围）内得以均衡分散。

3.2.1 对比、正交对比

记 n 个观测值（或称试验结果）为 $y_1,y_2,\cdots,y_n$。表达式
$$A=l_1y_1+l_2y_2+\cdots+l_ny_n$$
称为 $y_1,y_2,\cdots,y_n$ 之间的一个线性对比，如果系数 l_i 满足条件
$$l_1+l_2+\cdots+l_n=0, \qquad l_1^2+l_2^2+\cdots+l_n^2>0。$$
设再有另一对比
$$B=k_1y_1+k_2y_2+\cdots+k_ny_n$$
若 $\sum_i l_i k_i = 0$，则称对比 A 和 B 是正交的。

现将正交表中的试验结果 y，按列的不同水平分组求和，并考虑这些分组和之间的对比。由于表的正交性，每一列的分组和之间的任一对比都和任意其他列分组和之间的任一对比对 y_i 来说是正交的。例如表 1 的 $L_9(3^4)$ 中"列"的分组和
$$(y_1+y_2+y_3), \quad (y_4+y_5+y_6), \quad (y_7+y_8+y_9)$$
之间的一个对比（也是诸 y_i 之间的一个对比）
$$(y_1+y_2+y_3)-2(y_4+y_5+y_6)+(y_7+y_8+y_9)$$

和"品种"列的分组和
$$(y_1+y_6+y_8), (y_2+y_4+y_9), (y_3+y_5+y_7)$$
之间的一个对比
$$(y_1+y_6+y_8)-(y_3+y_5+y_7)$$
对 y_i 而言是正交的。容易验证，前一对比的系数按 $y_1,y_2,\cdots,y_9$ 的顺序是
$$1,1,1,-2,-2,-2,1,1,1$$
而后一对比的系数是
$$1,0,-1,0,-1,1,-1,1,0$$
故
$$\sum_i l_i k_i = 1\times 1+1\times 0+1\times(-1)+(-2)\times 0+(-2)\times(-1)$$
$$+(-2)\times 1+1\times(-1)+1\times 1+1\times 0=0。$$

应用上，常把对比中的系数 l_i 化为 $l_i/\sqrt{\sum l_i^2}$，以使系数平方和为 1。

3.2.2 正交与独立

设 y_i 是来自正态总体、方差为 σ^2 的独立观测值。对比 A 的方差为 $\sum l_i^2 \sigma^2$，对比 B 的方差为 $\sum k_i^2 \sigma^2$，而 A 与 B 的协方差为 $(\sum l_i k_i)\sigma^2$，故 A 与 B 之间的相关系数 r 由下式给出：

$$r^2 = \frac{(\sum l_i k_i)^2}{(\sum l_i^2)(\sum k_i^2)}。$$

当 A 与 B 正交，即 $\sum l_i k_i = 0$ 时，A 与 B 独立。因此，若用 A 估计某一参数，用 B 估计另一参数，则两个估计量将彼此独立。因各种因素效应（见后）是用正交表的按列分组和的对比去估计的，故不同因素效应的估计量之间是互相独立的。

设 $T_1, T_2, \cdots, T_s$ 为一列的 s 个分组和，$T = T_1+T_2+\cdots+T_s$ 为全部观测值和，则该列的（偏差）平方和为

$$ss(列)=(T_1^2+T_2^2+\cdots+T_s^2)/n-T^2/sn$$

其中 n 为分组和中的观测值个数。又设 $A_1, A_2, \cdots, A_{s-1}$ 是该列分组和的 $s-1$ 个正交对比，其系数平方和为 1，可以证明列平方和为

$$ss(列)=A_1^2+A_2^2+\cdots+A_{s-1}^2$$

以上公式在正交表的方差分析中是最基本的。

4. 析因试验及其部分实施

4.1 析因试验

在农业试验中,常考察不同品种、不同施肥、不同栽植密度等对产量的影响。这里,品种、肥料、密度等代表试验中的因素;不同品种、不同施肥量等代表有关因素的不同水平。类似地,在工业试验中,常考察温度、时间、压力、配方等因素对试验结果的影响。

在参试的诸因素中,把每一因素的每一水平同其余因素的每一水平组合起来进行试验,其目的为分析参试的诸因素对试验结果的影响,这样的试验叫做析因试验。设有 k 个因素 $f_1, f_2, \cdots, f_k$,各有 $s_1, s_2, \cdots, s_k$ 个水平,则共有 $s_1 \times s_2 \times \cdots \times s_k$ 个水平组合(又称处理组合)。如不作重复试验,则水平组合数就代表试验次数。不重复或等重复(每一水平组合重复同样多次)的析因试验,简记为 $s_1 \times s_2 \times \cdots \times s_k$ 型析因,显然,$s_1 \times s_2 \times \cdots \times s_k$ 型析因在试验范围内因素之间的配搭是均衡的,因而也是正交的。析因试验有两大好处。

第一,自然规律表明,为了得到好的试验结果,往往无法单独决定一个因素的哪一水平好,而要同时决定诸因素的哪一水平组合好,也就是说,要考虑交互作用。析因试验可以考察因素之间的交互作用。

第二,在析因试验中即使不作重复试验,对每个因素来说,由于同一水平的多次出现而包含着"隐藏的重复"。正是由于这种隐藏的重复有可能减少真正的重复(指同一水平组合的重复试验),甚至还可以考虑只对一部分水平组合进行试验,以减少试验次数。

4.1.1 主效应

在析因或在满足一定正交性(见后)的部分析因中,一个因素的水平变化所引起的试验结果的平均变化(比如前文谈过的两台机器的平均加工合格率的差异),可以认为主要是来自该因素本身的、不依赖于其他因素的水平变化的一种效应,称为该因素的主效应[①]。

4.1.2 交互作用

在析因试验中,称二因素存在交互作用(或称交互效应),是指其中一个因素的水平改变所引起的试验结果的改变和另一因素处在什么水平有

---

① 一些教科书中,又将主效应称零级交互效应,二因素的交互效应称一级交互效应。

关。例如,考虑因素 A、B 各取一、十两个水平的一个 2×2 或 2^2 析因,并假定 4 个水平组合(一,一),(十,一),(一,十),(十,十)的试验结果可以用数量 y_1, y_2, y_3, y_4 表示,如表 3 所示。

表 3

试验号	水平组合		交互作用	试验结果
	A	B	AB	
1	−	−	+	y_1
2	+	−	−	y_2
3	−	+	−	y_3
4	+	+	+	y_4

当 B 处于一或十水平时,随着 A 的水平从一变到十,试验结果的改变量分别是 (y_2-y_1),(y_4-y_3)。暂且不计试验误差,这两个改变量的差异

$$2AB = (y_4 - y_3) - (y_2 - y_1) = y_1 - y_2 - y_3 + y_4 \tag{1}$$

就代表 A 和 B 的交互作用(通常取其 1/2),记为 AB。如果这个差异等于零,则表示试验结果的改变量和 B 的水平无关,就说无交互作用,不难看出,在交互作用 AB 中,A 和 B 是对称的。值得注意,式(1)中的正负号正好是表 3 中 AB 列的符号。

三因素交互作用是指其中的二因素交互作用的大小与第三因素处在什么水平有关。多因素交互作用的涵义,依此类推。很明显,因素之间是否存在交互作用,其大小如何,及与因素所取的实际水平(指 0,1 或 2 所代表的有实际物理意义的水平)及水平变化范围有关。一般而言,水平数多或实际水平间隔大,试验结果在该范围的变化就大,出现交互作用的可能性也大,或者说,会出现较大的交互作用。在实际应用中,还要考虑随机误差。例如,当 $y_1 - y_2 - y_3 + y_4$ 虽不为零但又较小时,常被认为是随机误差所致,并不一定代表交互作用。

为方便起见,又称一个因素的主效应为一级效应;二个因素的交互作用为二级效应;三个因素的交互作用为三级效应;等等。

4.2 利用正交表作部分实施

当因素较多时,析因的全部水平组合数可能很大,考虑如何省去一部分水平组合,即所谓部分析因试验,从而减少试验次数,无疑有很大的现实意义。通常要求在一个部分析因试验中仍能有效地分析所需考察的因素的各级效应,特别是因素的主效应、二级效应以及个别必须考察的高级效

应。在某些情况下,通过正交设计,特别是正交表这个形式,可以方便地解决这个问题。

例如,在正交表 $L_4(2^3)$ 中可安排 $k=2$ 或 3 个各有两个水平的因素。当 $k=2$ 时,为 2^2 析因,可以把因素排在任意二列上,空出的一列为交互作用列。在无交互作用的情况下,可排 3 个因素,每列一个。2^3 析因本来有 8 个水平组合,现利用 $L_4(2^3)$ 只选出其中 4 个进行试验,故称为 2^3 析因的 1/2 实施。这 4 个水平组合由表 3 中的 4 行代号表示出来,每行代表一个水平组合。

一般地,当 s 为素数或素数幂时,可以通过适当的正交表获得 s^k 析因的 $1/s^r$ 实施(又记 s^{k-r} 析因),这里 r 是某个正整数。例如正交表 $L_9(3^4)$ 可用于 3^2 析因,或 3^3 析因的 1/3 实施,或 3^4 析因的 $1/3^2$ 实施(详见后)。

4.3 称重设计

在不计误差的情况下,两件东西合在一起称重等于它们分开称重之和。即一件东西的重量不受另一件东西是否合在一起称的影响,或者说,在称重问题上无交互作用。现在要在天秤上称出三件东西 A、B、C 的重量。一个自然的做法是:第一次先校正天秤,然后每物各称一次,共称 4 次。假定每称一次都有独立同分布的称重误差,且其标准差为 σ。由于每件物重的估计量将是两次读数之差:一次是校正天秤时的读数,另一次是该物称重时的读数,故它的标准差为 $\sigma\sqrt{2}$。一个改进的方法是,除校正天秤一次外,每二件称一次,即 A 和 B,A 和 C,B 和 C 合起来各称一次,总共也是称 4 次。可把 A、B、C 看作因素,并利用正交表 $L_4(2^3)$ 说明这个改进的方法相当于一个 2^3 析因的 1/2 实施,如表 4 所示。表中的正负号(相当于前表中的 0 和 1)代表因素的两个水平。"—"表示不放在天秤上,"+"表示放在天秤上。假定 4 次称重的读数依次是 y_1、y_2、y_3、y_4,A 的重量应估计为

$$\frac{1}{2}(-y_1+y_2-y_3+y_4)=\frac{1}{2}(y_2+y_4)-\frac{1}{2}(y_1+y_3) \qquad (2)$$

(注意,这个估计方法本身说明无须另校正天秤。)估计的标准误差是 σ。至于 B 和 C 的重量估计,按表中的 +、— 号在式(2)中做适当的变化即可。它们的标准差当然也是 σ。

表 4

$L_4(2^3)$

称重序号	A	B	C	重量
1	−	−	−	y_1
2	+	−	+	y_2
3	−	+	+	y_3
4	+	+	−	y_4

更好的方法是对两个水平记号作另外的解释。"−"表示放在天秤左边,"+"表示放在天秤右边。这样 4 次称重,每次都称三件,只是看哪一件放在天秤的哪一边罢了。A 的重量的估计式仍然和(2)式类似,只须把式中系数 1/2 改为 1/4。但这样一来,估计量的标准差将进一步减为 $\sigma/2$。在称重次数为 4 的条件下,这是最优的一种称重方法。通常所说的称重设计是指后一种设计。一般地,若有 k 件称重物品,可以使用二水平正交表在天秤上设计 n 次组合称重,使其估计量的标准差为 $\sigma/\sqrt{n}$,但这里 n 必须大于或等于 $k+1$ 且能被 4 除尽。

"称重设计"说明部分析因的一个理想情形:"因素"之间无交互作用,同时也说明,析因试验的隐藏重复性使"效应"估计量的误差得以下降。

4.4 正交表试验的直观分析

正交表主要用于部分析因试验的设计与分析。全面析因试验则无须使用正交表,其分析也将与方差分析类同(见后)。用正交表安排试验,首先要了解有哪些常用的正交表,然后对定好的因素个数和水平数选用合适的正交表。

常用正交表有大小不一的二水平、三水平、四水平、二与四混合水平、五水平等正交表,见附录。正交表的行数代表试验次数,列数则代表可容纳的因素最大个数。在实际应用中,用表大小常有可选择的余地。小表可以节省试验次数,大表则有利于考察交互作用和试验误差。

例 1:为了提高"510"硅油(一种脱模剂)的黏度,准备通过试验来考察聚合温度、聚合时间、搅拌速度和真空压力等四个因素对硅油黏度的影响,从而找出较好的工艺条件。根据专业知识,对每个因素各取两个水平如表 5 所示。

表 5

水平	A 聚合时间（分钟）	B 聚合温度（℃）	C 搅拌速度	D 真空压力（kg/mm）
0	135	84	慢	680
1	165	77	快	710

全面析因试验要对 $2^4=16$ 个不同的水平组合一一进行试验。现利用正交表 $L_8(2^7)$（见附录）选做 8 个。试验的设计与分析，如表 6 所示。

表 6

试验号	1 聚合时间 A	2 聚合温度 B	3 CD AB	4 搅拌速度 C	5 BD AC	6 AD BC	7 真空压力 D ABC	黏度
1	0(135′)	0(84 ℃)	0	0(慢)	0	0	0(680)	1384.0
2	1(165′)	0	1	0	1	0	1(710)	1143.0
3	0	1(77 ℃)	1	0	0	1	1	959.8
4	1	1	0	0	1	1	0	797.9
5	0	0	0	1(快)	1	1	0	2927.0
6	1	0	1	1	0	1	0	3570.0
7	0	1	1	1	1	0	0	2000.0
8	1	1	0	1	0	0	1	2288.0
m_0	1817.7	2256.0	1849.2	1071.1	2050.5	1703.7	1937.9	
m_1	1949.7	1511.4	1918.2	2696.3	1716.9	2063.7	1829.5	
m_1-m_0	132.0	−744.6	79.0	1625.2	−333.6	360.0	−108.4	

这里，把聚合时间、聚合温度、搅拌速度和真空压力分别排在表的第 1、2、4、7 列上。这样的安排，在三因素以上的交互作用可忽略的假定下，便可考察主效应和某些二因素交互作用。为了考察这些效应，在表下端算出 m_0，m_1 和 m_1-m_0，其中 m_0 一栏是把各列中的"0"水平所对应的 4 次试验结果相加，再除以 4（取平均）。例如，B 列的 m_0 是把 B 列中"0"水平（84 ℃）所对

应的第 1,2,5,6 次试验结果相加,而后取平均,即

$$B \text{ 列的 } m_0 = \frac{1}{4}(1384.0+1143.0+2927.0+3570.0)=2256.0$$

其他列的 m_0 的计算照此类推。m_1 的计算也完全相仿,只不过把"0"水平换为"1"水平罢了。至于 m_1-m_0,意义更为明显,例如,

$$A \text{ 列的 } m_1-m_0=1949.7-1817.7=132.0。$$

从 m_0、m_1 和 m_1-m_0 的计算过程可以看出,对于排有因素的列来说,m_0 是所排因素取 0 水平时的 4 次试验结果的平均值;m_1 是取 1 水平时另 4 次试验结果的平均值。因此,绝对值 $|m_1-m_0|$ 就是该因素的水平变动所引起的指标值的平均变化幅度或极差,它反映了该因素的主效应。主效应越大,说明因素在所考察的水平变化范围内对指标的影响越大。因此,可按 $|m_1-m_0|$ 的大小排出诸因素的主次关系。

比较各列 m_1-m_0 的绝对值,可以看到,搅拌速度 C 列的 $|m_1-m_0|=1625.2$ 最大。其次是 B 列的 $|m_1-m_0|=744.6$。A 列和 D 列的 $|m_1-m_0|$ 均较小。故搅拌速度是最主要因素,聚合温度次之,聚合时间和真空压力再次之。因为黏度越大越好,对于搅拌速度,$m_1=2696.3$,$m_0=1071.1$,故取其 1 水平,即快速搅拌比慢速搅拌要好。同理,对于聚合温度,取其 0 水平即 84 ℃。对于更次要的因素,由于它在所考虑的水平变化范围内,对指标的影响微小,故其水平一般可以任意确定一个,或者根据其他要求(如节省原料、运费,方便操作等)来决定。

"510"硅油后来的试验表明,只要抓住了搅拌速度这个主要因素,采用快速搅拌,在 84 ℃下进行裂解聚合反应,黏度可高达 6000 以上。

表 6 中第 6 列的 $|m_1-m_0|=360$,其大小居第三位。根据表的构造原理(见后),第 6 列代表交互作用 BC 或 AD。由于 C 的主效应大,故认为此列主要代表交互作用 BC,而不代表 AD。这样,就要考虑因素 B 和 C 的水平搭配关系。可以通过列表的方法来比较 B 和 C 的各种水平搭配所对应的平均黏度(见表 7)。从表中看到,B_0C_1 这种水平搭配所对应的平均黏度最高,即采取高温下快速搅拌这一工艺条件较好。这和按主效应大小分析的结果是一致的。当然,有时也会不一致,那就要做进一步的试验和分析了。

表 7
搅拌

		慢(C_0)	快(C_1)
温度	84 ℃(E_0)	$\frac{1384+1143}{2}=1263.5$	$\frac{2927+3570}{2}=3248.5$
	77 ℃(B_1)	$\frac{959.8+797.9}{2}=878.8$	$\frac{2000+2288}{2}=2144$

上例是对单指标(黏度)的直观分析,下面再考虑多指标情形。

在实际应用中常遇到多指标的问题。多指标的试验分析,虽然比单指标的情形复杂一些,但基本上是重复单指标的计算分析方法,即先对每一个指标逐一地进行计算,然后综合分析,选取能兼顾各个指标的较好水平组合。

例 2:橡胶配方试验。目的是要改进能反映橡胶性能的三个指标值:伸长率越大越好;变形越小越好;屈曲次数越多越好。参试因素及其水平的制定如表 8 所示。

表 8

水平	促进剂总量 A	氧化锌总量 B	促进剂甲 所占比例 C(%)	促进剂乙 所占比例 D(%)
1	2.9	1	30	34.7
2	3.1	3	25	39.7
3	3.3	5	35	44.7
4	3.5	7	40	49.7

为了节约试验次数,不考虑因素之间的交互作用。利用四水平正交表 $L_{16}(4^5)$ 安排试验。这是一个 4^4 析因的 1/4 实施。试验结果及其分析见表 9 和表 10。

对每一个指标,表中的 $m_i(i=1,2,3,4)$ 都是对应于第 i 水平的试验结果的平均值。例如,对于伸长率:

B 列的 $m_4=(505+480+515+475)/4=494$;

对于屈曲:C 列的 $m_2=(3.9+3.2+2.9+2.3)/4=3.1$;

其余类推。

表 10 最后一行"极差"是各列 $m_i(i=1,2,3,4)$ 中最大值与最小值之差。例如,对于变形,

A 列最大值减最小值 $=47-44=3$；

对于伸长率，M 列最大值减最小值 $=513-504=9$；

其余类推。

这些极差的大小反应了因素（主）效应的大小。现对每一指标，将诸因素按其效应的大小分别主次排列如下（效应小的可加以忽略）：

$$\text{主}\longrightarrow\text{次}$$

伸长率：A　B　C

变　形：C　A　B

屈　曲：A　B

表 9

试验号	$L_{16}(4^5)$					伸长率 (%)	变形 (%)	屈曲 (万次)
	1 A	2 B	3 C	4 D	5 "空"			
1	1(2.9)	1(1)	1(30)	1(34.7)	1	545	40	5.0
2	1	2(3)	2(25)	2(39.7)	2	490	46	3.9
3	1	3(5)	3(35)	3(44.7)	3	515	45	4.4
4	1	4(7)	4(40)	4(49.7)	4	505	45	4.7
5	2(3.1)	1	2	3	4	492	46	3.2
6	2	2	1	4	3	485	45	2.5
7	2	3	4	1	2	499	49	1.7
8	2	4	3	2	1	480	45	2.0
9	3(3.3)	1	3	4	2	566	49	3.6
10	3	2	4	3	1	539	49	2.7
11	3	3	1	2	4	511	42	2.7
12	3	4	2	1	3	515	45	2.9
13	4(3.5)	1	4	2	3	535	49	2.7
14	4	2	3	1	4	488	49	2.3
15	4	3	2	4	1	495	49	2.3
16	4	4	1	3	2	475	42	3.3

表 10

m_i 值	伸长率					变形					屈曲				
	A	B	C	D	"空"	A	B	C	D	"空"	A	B	C	D	"空"
m_1	514	534	504	512	515	44	46	42	46	46	4.5	3.6	3.4	3.0	3.0
m_2	489	501	498	504	508	46	47	47	46	47	2.4	2.9	3.1	2.8	3.1
m_3	533	505	512	506	512	46	46	47	46	46	3.0	2.8	3.1	3.4	3.1
m_4	498	494	519	513	499	47	44	48	46	46	2.7	3.2	3.0	3.3	3.2
极差	44	40	21	9	16	3	3	6	0	1	2.1	0.8	0.4	0.5	0.2

在多指标的分析中，常常会碰到这样的情况：对不同的指标可能筛选出不同的重要因素以及不同的最优水平组合，这时，如何决定最优水平组合，就需要根据各个指标的重要性，权衡得失，统筹兼顾。在本例中，因素 A 对伸长率和屈曲两个指标的影响最大。按屈曲，A 取 A_1 最好；依伸长率，A 取 A_3 或 A_1 最好。A 在变形指标中居第三位，从变形看，A 取 A_1 好，综合三个指标 A 取 A_1。类似地分析其他因素，决定 B 取 B_1，C 取 C_1。D 对每个指标的影响都不大，其水平选取，可以灵活。

所选的水平组合 $A_1B_1C_1$（D 任意）和表中第一号试验条件基本相符，其试验结果是比较好的。

表中计算了"空"列的极差，可作为误差来参考使用。① 这样，对于伸长率，A 列和 B 列的极差相对于误差而言是比较大的；对于变形，A、B、C 三个因素列的极差都是比较大的，尤其是 C 列；对于屈曲，A 列和 B 列的极差也都明显地超出误差范围。

正交表试验的直观分析是相对于方差分析而言。方差分析要求有较严格的试验误差方差和因素效应方差的计算，从而能作出关于每一效应的显著性判断（见方差分析篇）。

5. 正交表中交互作用列的表示与构造

正交表按其构造特性可分为两类：一类是具备有各级交互作用列的正交表，可用来分析在析因试验中因素的各级效应；另一类是不具备交互作用列或只具备个别交互作用列的正交表。如果用这类表安排部分析因，则

---

① 应理解为有所参考比没有好，直观分析不要求严格论证。

一般只能在无交互作用的假定下分析因素的主效应。①

本文此处只限于讨论第一类表（第二类表将在后文提到它的用处），为简单起见，限于讨论各因素的水平数均相同且为素数 p 的情形（当水平数为素数幂时也能同样讨论，但在表的构造上须引进有限域的概念）。它包括了最常用的二水平正交表

$$L_{2^t}(2^k), 其中 k=2^t-1$$

和三水平正交表

$$L_{3^t}(3^k), 其中 k=(3^t-1)/2 \quad (t=2,3,4,\cdots)。$$

这类表的主要构造特性是，不但各级交互作用分别反映在表的不同列中，而且同一级的交互作用还按其所含自由度的多少而进一步（正交地）分解在不同的列上。

5.1 p^k 析因的自由度分解

k 个因素各 p 个水平的析因试验的方差分析表给出各级效应（交互作用）的自由度如表 11 所示。

表 11

各级效应	自由度	因素组合数
每个因素的主效应	$p-1$	C_k^1
每两个因素的交互作用	$(p-1)^2$	C_k^2
…		
每 $k-1$ 个因素的交互作用	$(p-1)^{k-1}$	C_k^{k-1}
k 个因素的交互作用	$(p-1)^k$	C_k^k

全部自由度为：

$$C_k^1(p-1)+C_k^2(p-1)^2+\cdots+C_k^{k-1}(p-1)^{k-1}+(p-1)^k$$
$$=[(p-1)+1]^k-1=p^k-1。$$

在 p 水平的正交表中，每列的自由度为 $p-1$。从而推知各级效应在表中所应占的列数。

当 $p=3$ 时，每列的自由度（d.f.）为 2。但每二因素的交互作用有 $(p-1)^2=2^2=4\,d.f.$，故须占表的 2 列。又因每三因素的交互作用有 $(p-1)^3=2^3=8\,d.f.$，故须占表的 4 列。如此类推。

又如，当 $p=5$ 时，每列的自由度为 $p-1=4$，二因素交互作用占 $(p-$

---

① 某些无交互作用列的正交表，如附表（三），仍提供有分析个别交互作用的可能性。

$1)^2 = 4^2 = 16 d.f.$，故须占表的 4 列。如此类推。

5.2 交互作用列的表示

对因素 $A, B, C, \cdots$ 用相同的单个字母记因素的主效应列；用两个字母的组合记二因素交互作用列；用三个字母的组合记三因素交互作用列等。按不同的水平数，具体表示如表 12 所示。

表 12　正交表列名

二水平：A　B　AB　C　AC　BC　ABC　D　…

三水平：A　B　AB　AB^2　C　AC　AC^2　BC　BC^2　ABC　AB^2C　ABC^2　AB^2C^2　D　…

五水平：A　B　AB　AB^2　AB^3　AB^4　C　AC　AC^2　AC^3　AC^4　BC　BC^2　BC^3　BC^4　ABC　ABC^2　ABC^3　ABC^4　…　AB^4C^4　D　…

从表中看到，当水平数 $p=2$ 时，每个效应（相当于因素 $A, B, C, \cdots$ 的每一种组合）各占一列。当 $p=3$ 时，每个一级效应各占一列；每个二级效应各占二列，比如 A 和 B 的交互作用占 AB、AB^2 二列；每个三级效应各占四列，比如 A、B、C 三者的交互作用占 ABC、AB^2C、ABC^2、AB^2C^2 四列。当 $p=5$ 时，每个一级效应各占一列；每个二级效应各占四列，比如 A 和 B 的交互作用占 AB、AB^2、AB^3、AB^4 四列。对每一种水平数来说，每列都有一个不同的字母表示的列名。不难看出这些字母表现的规律。例如，对 $p=3$ 水平情形，每增加一个新字母（表示增加一个因素），便用这个字母的 1 次方和 2 次方直至 $p-1$ 次方依次乘前面的各列，从而构成随后的各列。注意，每个列名的字母都按字典顺序排，并且每个列名的第一个字母都是 1 次方，以避免表示形式上的重复。通常把这种列名叫作标准化列名。

5.3 列名与水平的对应关系

把用单个字母表示的列叫做基本列。根据这类正交表的构造理论，对于水平数为素数 p 的情形，给定基本列的水平，就能按下述模 p 加法写出各交互作用列的相应水平。基本列 $A, B, C, \cdots$ 的水平 $x_1, x_2, x_3, \cdots$，对于 $i=1, 2, \cdots, x_i = 0, 1, \cdots$ 或 $p-1$，应构成析因中的全部水平组合。这些组合可按某种顺序比如按字典法排列出来。

水平和列名的对应关系，按不同水平情形给出，如表 13 所示。

表 13　二水平正交表

列名	A	B	AB	C	AC	BC	ABC	⋯
水平	x_1	x_2	x_1+x_2	x_3	x_1+x_3	x_2+x_3	$x_1+x_2+x_3$	⋯

基本列 $A,B,C,\cdots$ 的水平为 0 或 1，取决于 2^k 析因中的特定水平组合，其余列的加法运算一律按模 2 化简。如当 (A,B,C) 的水平组合为 $(1,0,1)$ 时，AB 列的相应水平为 $1+0=1$；AC 列的相应水平为 $1+1=2\equiv 0$(模 2)，等。参考附录的二水平正交表。

$A,B,C,\cdots$ 列的水平为 0,1 或 2，决定于 3^k 析因中的特定水平组合。其余列按模 3 加法计算。例如，对于 (A,B,C) 的特定水平组合 $(2,1,2)$，AB^2C^2 的相应水平为 $2+2\times 1+2\times 2=8\equiv 2^3$(模 3)。如表 14 三水平正交表所示。

表 14　三水平正交表

列名	A	B	AB	AB^2	C	AC	AC^2	BC
水平	x_1	x_2	x_1+x_2	x_1+2x_2	x_3	x_1+x_3	x_2+2x_3	x_2+x_3
列名	BC^2	ABC	ABC^2	AB^2C	AB^2C^2	⋯		
水平	x_2+2x_3	$x_1+x_2+x_3$	$x_1+x_2+2x_3$	$x_1+2x_2+x_3$	$x_1+2x_2+2x_3$	⋯		

五水平情形与此相仿。例如，对于 A,B,C 的水平 x_1,x_2,x_3，AC^3 列的水平为 x_1+3x_3(模 5)；AB^4C^4 的水平为 $x_1+4x_2+4x_3$(模 5)等。

5.4　广义交互作用列

按上述方法所构造的正交表，实质上是由它的基本列所决定，其余各列均为由基本列所产生的各级交互作用列。现在的问题是，若从表中任取一些列，将它们看成基本列，而后按上述方法构造它们的交互作用列，是否可以证明这些列依然是表中的列？答案是肯定的。正因为如此，我们把这类表称为完备型正交表，而把任何二列或多列所产生的列称为它们的广义交互作用列。由于完备型正交表具有这一性质，在做试验设计时，不管把因素排在表中的哪些列上，都能在表中找到它们的各级交互作用列。

下面按不同的水平数分述广义交互作用列的求法。

5.4.1　二水平情形

求二列的广义交互作用列时，将其列名按代数乘法规则相乘，然后将字母的指数按模 2 化简。例如，AB 列和 BC 列的广义交互作用列为

$$AB\cdot BC=AB^2C=AC,$$

就是说,它们的广义交互作用列在 AC 列上。

容易验证,在正交表 $L_4(2^3)$ 中,任二列的广义交互作用列是剩下的那一列。

5.4.2 三水平情形

求二列的广义交互作用列时,将其中一列及其平方乘另一列,从而得到两个交互作用列。然后将字母的指数按模 3 化简。如果所得乘积的第一个字母不是 1 次方,则取这个乘积的平方,将其化为标准化列名。例如,AB 和 AB^2C 的广义交互作用列是

$$AB \cdot AB^2C = A^2B^3C = A^2C = (A^2C)^2 = AC^2$$

和

$$AB \cdot (AB^2C)^2 = A^3B^2C^2 = B^2C^2 = (B^2C^2)^2 = BC。$$

容易验证,在正交表 $L_9(3^4)$ 中,任二列的广义交互作用列是其余二列。

5.4.3 五水平情形

求二列的广义交互作用列时,将其中一列及其平方、立方和 4 次方分别乘另一列而得到 4 个广义交互作用列。再将字母的指数按模 5 化简。如果所得乘积的第一个字母不是 1 次方,则可通过取该乘积的平方或立方或 4 次方,将其第一个字母化为 1 次方而得到标准化列名。

不管是几个水平的情形,三列的广义交互作用列可通过先求出其中二列的广义交互作用列后,再求这些交互作用列与第三列的广义交互作用列得之。更多列的广义交互作用列的求法由此类推。(有可能求出的结果是 I,如对于二水平情形,A,BC 和 ABC 三列的广义交互作用列是 $A \cdot BC \cdot ABC = I$,这说明这三列互为交互作用列。同时 I 也可解释为表中的一个平均效应列,但这列不用来安排任何参试因素,因此它不出现在通常的正交表里。)

5.5 正交表的等价变换

同一张正交表可以变为许多不同的形式,例如,通过如下的初等变换:

(A) 行的置换,即行次序的重新排列(这相当于试验序号的改变);

(B) 列的置换,即列次序的重新排列(这相当于因素记号的改变);

(C) 同一列内号码的置换,即不同号码的一次或多次对调(这相当于把同一因素的不同水平记号互换)。

而将一张正交表变到另一张正交表,凡是通过以上三种初等变换而得到的正交表都是等价的,即它们没有实质上的差异。

不消说,表中的号码 0,1,2,$\cdots$,不过是区分若干个不同水平的记号而已,完全可以代之以另外一些号码或符号如 1,2,3,$\cdots$,或甲,乙,丙,$\cdots$,等

等。参看表 15。

表 15　$L_9(3^4)$ 的等价变换

0	0	0	0	1	1	1	1	1	1	1	1	1	1	1	1
1	0	1	1	2	1	2	2	1	2	2	2	1	2	2	3
2	0	2	2	3	1	3	3	1	3	3	3	1	3	3	2
0	1	1	2	1	2	2	3	2	1	2	3	2	1	2	2
1	1	2	0	2	2	3	1	2	2	3	1	2	2	3	1
2	1	0	1	3	2	1	2	2	3	1	2	2	3	1	3
0	2	2	1	1	3	3	2	3	1	3	2	3	1	3	3
1	2	0	2	2	3	1	3	3	2	1	3	3	2	1	2
2	2	1	0	3	3	2	1	3	3	2	1	3	3	2	1

变换关系：$\{0\to1, 1\to2, 2\to3\}$；第一、二列对调；第四列 2、3 对调。

6. 部分析因中的效应混杂

6.1　效应混杂

析因试验的部分实施可以通过列举析因中的一部分水平组合来表示，但一个更为简便的方法是，选择适当的正交表，把一些因素首先排满基本列，再把其余的因素排在交互作用列上，然后根据这些排有因素的列的水平组合进行试验。但究竟把哪些因素排在哪些交互作用列上，才不致影响对希望考虑的各级效应的分析？这就需要弄清楚部分实施中的效应混杂关系。例如，在表 $L_4(2^3)$ 中做 2^3 析因的 1/2 实施，如表 16 所示，由于因素 C 排在 AB 列上，C 的主效应将与 AB 的交互作用相混杂，就是说，两者表现在同一列上。不仅如此，由于表中任二列的广义交互作用列是其余一列，若用等号"="表示"与之混杂"，则

$$\begin{cases} A = BC, \\ B = AC, \\ C = AB。 \end{cases} \tag{3}$$

表 16

A BC	B AC	C AB
0	0	0
1	0	1
0	1	1
1	1	0

同理,由于表 $L_9(3^4)$ 中任二列的广义交互作用列是另外二列,如果除 A、B 列外,还把因素 C 排在 AB 列上,如表 17 所示,则有如下混杂关系:
$$A=BC^2, \quad B=AC^2, \quad C=AB,$$
$$AB^2=AC=BC, \quad 等等。$$

表 17

A BC^2	B	C AB AC^2	AB^2 AC BC
0	0	0	0
1	0	1	1
2	0	2	2
0	1	1	2
1	1	2	0
2	1	0	1
0	2	2	1
1	2	0	2
2	2	1	0

因此,部分实施的设计要求,对于希望考察的各级效应不要与实际存在的效应相混杂。

6.2 定义方程

部分析因中的全部效应混杂关系可以用一个叫作定义方程的恒等关系式来描述。这个恒等关系式同时也区分了析因中的哪些水平组合被包括、哪些水平组合不被包括在这个部分析因中;特别是对于二水平情形,在被包括和不被包括的水平组合之间形成一个或多个对比,因此这个恒等关系式又叫作定义对比。

6.2.1 二水平情形

在 $L_4(2^3)$ 中把 C 排在 AB 列上时(见表 17),将出现混杂 $C=AB$。若两边乘以 C,则得 $I=ABC$,这里按照习惯把 I 写成恒等符号 I。称 ABC 为定义效应,并称这个恒等关系式为定义方程。这个定义方程概括了 2^3 析因的 1/2 实施中的所有混杂关系[式(3)]。例如,要找出 A 的混杂关系,用

A 乘定义方程两边便得：$A=BC$。[$I=ABC$ 同时也表示只挑选使 ABC 取同一水平的那些水平组合进行试验,参看正交表 $L_8(2^7)$。]

假定在正交表 $L_8(2^7)$ 上排 5 个因素 $A、B、C、D、E$ 做 1/4 实施。除 $A、B、C$ 占三个基本列外,把 D 和 E 分别排在 ABC 和 BC 两列上即有混杂关系：

$$D=ABC \quad 和 \quad E=BC,$$

分别乘以 D 和 E 后,又可写成

$$I=ABCD \quad 和 \quad I=BCE,$$

其中 $ABCD$ 和 BCE 代表定义效应。因

$$ABCD \cdot BCE = ADE,$$

故又有 $I=ADE$。从而得到这个 1/4 实施的定义方程(组)：

$$I=ABCD=BCE=ADE。 \tag{4}$$

它概括了全部的混杂关系。例如,用 AB 去遍乘这个方程(组)就得到 AB 的全部混杂关系：

$$AB=CD=ACE=BDE。$$

一般,二水平析因的 $1/2^r$ 实施的定义方程可以写成如下形式：

$$I=X=Y=XY=Z=XZ=YZ=XYZ=U=\cdots$$

其中 $X,Y,Z,U,\cdots$,代表 r 个定义效应。

6.2.2 三水平情形

考虑 3^3 析因的 1/3 实施。除 A,B 排在 $L_9(3^4)$ 的基本列外,把 C 排在 AB 列上将有混杂关系 $C=AB$。

两端乘以 C^2 得(指数按模 3 化简)：

$$I=ABC^2,$$

其中 ABC^2 为定义效应,恒等关系式代表定义方程。它概括了这个 1/3 实施的全部混杂关系。为了求某一效应的混杂关系,可对定义效应补充一个平方项,将式(4)扩大为

$$I=ABC^2=(ABC^2)^2, \tag{5}$$

然后将该效应遍乘这个扩大了的方程式(5)即得。例如,将 A 乘式(5)得

$$A \times I = A \times ABC^2 = A \times (ABC^2)^2,$$

再将指数按模 3 化简,并使第一个字母为 1 次方得

$$A=AB^2C=BC^2。$$

这说明主效应 A 不仅与二级效应 BC^2 混杂,而且与三级效应 AB^2C 混杂。

又假定在表 $L_{27}(3^{13})$ 中作 1/9 实施。除因素 A,B,C 占三个基本列外,

把因素 D 和 E 分别排在 ABC 和 AB^2 列上,则有
$$I=ABCD^2 \quad 和 \quad I=AB^2E^2。$$
根据三水平广义交互作用的求法,
$$ABCD^2 \cdot AB^2E^2 = A^2CD^2E^2 = (A^2CD^2E^2)^2 = AC^2DE$$
$$ABCD^2 \cdot (AB^2E^2)^2 = B^2CD^2E = (B^2CD^2E)^2 = BC^2DE$$
故又有
$$I=AC^2DE \quad 和 \quad I=BC^2DE^2$$
由此得这个 1/9 实施的定义方程
$$I=ABCD^2=AB^2E^2=AC^2DE=BC^2DE^2。$$

为了求某一效应的混杂关系,仍然要先对这个方程中的每一效应(包括定义效应及其广义交互作用)配上平方项,再用代表该效应的字母组合去遍乘扩大了的定义方程。

一般,三水平析因的 $1/3^r$ 实施的定义方程有如下形式:
$$I=X=Y=XY=XY^2=Z=XZ=XZ^2=YZ=YZ^2$$
$$=XYZ=XYZ^2=XY^2Z=XY^2Z^2=U=\cdots$$
其中 $X,Y,Z,U,\cdots$ 表示 r 个定义效应。

6.3 应用——R 分解方案

通常,在析因试验中,只考虑一级效应(主效应)和二级效应(指二因素交互作用),而常假定三级和三级以上的效应为零。例如,从定义方程(4)可以得到
$$CD=AB=BDE=ACE$$
若假定 BDE 与 ACE 为零,则只有二级效应这一部分被混杂:$CD=AB$。

当定义方程中的每一定义效应及其广义交互作用都由至少 R 个字母组成时,它所定义的部分实施是一个 R 分解方案。当 $R=5$ 时,称为 V 分解方案。这时用任何一个或二个字母去乘定义方程,其结果都不会少于 3 个字母,从而知道一、二级效应只会同三级以上(包括三级)的效应相混杂。既然认为三级以上的效应为零,那么一、二级效应就算没有被混杂。

因此,一个部分析因试验最好是 V 分解方案,以使一、二级效应不至于混杂。但有时为了能作试验次数较少的部分实施而达不到 V 分解的要求,则必须混杂某些一、二级效应。此时应选择一个部分实施方案,使被混杂的一、二级效应仅是那些较为次要的效应。

例如,假定要在 $L_{16}(2^{15})$ 上安排 6 个因素作 1/4 实施,比较下面两种

方案：

(1) 把 E 和 F 分别排在 ABC 和 BCD 列上，从而得定义方程
$$I = ABCE = BCDF = ADEF$$

(2) 把 E 和 F 分别排在 $ABCD$ 和 BCD 列上，从而得定义方程
$$I = ABCDE = BCDF = AEF$$

可以认为第(1)种方案比第(2)种方案好。因为前者是 Ⅳ 分解方案，只有二级效应互相混杂；而后者是 Ⅲ 分解方案，不免有某些主效应与二级效应混杂。

7. 区组混杂

当析因或其部分实施中的水平组合较多而不能同时放在一个"完全区组"中进行试验时，需要把这些水平组合分为若干部分，分别放到若干个"不完全区组"中进行试验。如何合理地划分这些水平组合，亦即如何划分不完全区组，正交表提供了简便的方法。现分述如下。

7.1 二水平情形

且把区组当作"因素"，排在正交表的某列上。其结果将使区组效应（不同区组的差异）与该列原来所代表的因素效应（或交互作用）互相混杂（一般不考虑区组与其他参试因素的交互作用）。①

以表 $L_8(2^7)$ 为例，在 A,B,C 三个基本列上排三个试验因素，在 ABC 列上排一个区组"因素" b_1。这样就把 ABC 列的 4 个 0 水平所对应的试验号，如表 17 的第 1,4,6,7 号划分为一个区组，且记为区组 Ⅰ。并把同列的 4 个 1 水平所对应的试验号，如表中的第 2,3,5,8 号划为另一个区组，且记为区组 Ⅱ。由于区组 Ⅰ 的试验结果（取其总和或平均）同区组 Ⅱ 的试验结果相比，不仅代表交互作用 ABC，而且代表两个区组之间的差异，即所谓区组效应，所以说 ABC 效应为区组效应所混杂。其他各列效应均不为区组所混杂。这是因为其他各列的两个水平都平分在不同区组之中，也就是说它们和区组正交。所以，在交互作用 ABC 无须考察的情况下，这样的区组划分是合理的。

---

① 文献中把因素效应之间的相互混杂现象叫做互为别名（aliasing），而把混杂（confounding）一词专门用于描述因素效应与区组之间的混杂。但在我国推广应用工作中很少使用别名一词。

表 17
$L_8(2^7)$

序号	A	B	AB	C	b_2 AC	BC	b_1 ABC	区组划分
1	0	0	0	0	0	0	0	I_0
2	1	0	1	0	1	0	1	II_1
3	0	1	1	0	0	1	1	II_0
4	1	1	0	0	1	1	0	I_1
5	0	0	0	1	1	1	1	II_1
6	1	0	1	1	0	1	0	I_0
7	0	1	1	1	1	0	0	I_1
8	1	1	0	1	0	0	1	II_0

如果再利用 AC 列安排第 2 个区组"因素"b_2,进而把水平组合按试验号分为(1,6),(4,7),(2,5),(3,8)4组,则不仅交互作用 AC 为区组所混杂,而且 ABC 和 AC 的广义交互作用 $ABC \cdot AC = B$ 也被区组混杂了。从表 17 可以看到,B 列的 0 水平对应着两个区组(I_0 和 II_1),而它的 1 水平又对应着另外两个区组(I_1 和 II_0),即区组差异和 B 的主效应混杂在一起了。

一般,若效应 $X,Y,Z,\cdots$,被区组混杂,则它们的各级广义交互作用也都被区组混杂。

顺便提一下,在区组混杂设计中,往往不考虑区组与区组之间以及区组与因素之间的交互作用。

7.1.1 部分析因中的区组混杂

在部分析因中,正交表的每一列都混杂有二个或多个效应。这时,若把区组因素排在表的某一列上,它自然要同所有混杂在该列上的效应都发生混杂。例如,在由定义方程

$$I = ABCD$$

给出的 2^4 析因的 1/2 实施中,再按 BC 列划分区组。从定义方程得

$$BC = AD,$$

即知 BC 和 AD 同为区组所混杂,故整个正交设计方案可表示为:

定义方程:$I = ABCD$;   区组混杂:$BC = AD$。

这里等号"="仍然表示"与之混杂"。为了找出某一效应的混杂关系,仍然是用代表该效应的字母去乘定义方程(但不去乘区组混杂方程;与区组相混杂的效应已全部表明在区组混杂方程之中)。

如果再用 AC 列安排第二个区组"因素",则设计方案可写成:

定义方程: $\quad\quad\quad I=ABCD$;

区组混杂: $\quad\quad BC=AD, AC=BD, AB=CD$。

最后一个等式是前三个等式的乘积,这点是容易理解的。

7.1.2 参考方案

以上设计方案仅为了说明概念而取其简单,并未考虑其实用性。在实用的区组混杂设计中,一般要求做到一、二级效应不被区组混杂,这相当于在区组混杂方程中的每一个效应均不少于 3 个字母。但有时为了能使用较小的区组以保证组内小区(泛指试验的操作单元)的匀一性,而不可避免地要混杂个别二级效应,这时可适当调换因素字母,使被混杂的二级效应尽可能是次要的。

下面列举一些较实用的设计方案供参考。

① 2^6 析因的 1/2 实施、划分为 4 个区组:

定义方程: $\quad\quad I=ABCDEF$;

区组混杂: $\quad\quad ACE=BDF, CD=ABEF,$
$\quad\quad\quad\quad\quad ADE=BCF。$

实现方法:在正交表 $L_{32}(2^{31})$ 中,除把 A、B、C、D、E 排在基本列上外,把 F 排在 $ABCDE$ 交互作用列上,并用 ACE 和 CD 二列划分区组,每区组包含 8 个水平组合。由于 4 个区组的划分,不可避免地要有一个二级效应被区组混杂,这里选为 CD。

② 2^8 析因的 1/4 实施、划分为 4 个区组:

定义方程: $\quad I=ABCDG=ABEFH=CDEFGH$;

区组混杂: $\quad ACF=BDFG=BCEH=ADEGH,$
$\quad\quad\quad\quad BDE=ACEG=ADFH=BCFGH,$
$\quad\quad\quad\quad ABCDEF=EFG=CDH=ABGH。$

实现方法:在正交表 $L_{64}(2^{63})$ 中,除把 A、B、C、D、E、F 排在基本列上外,把 G 和 H 分别排在 $ABCD$ 和 $ABEF$ 二列上。另外,再利用 ACF 和 BDE 二列将 64 个水平组合划分为 4 个不完全区组。

③ 2^8 析因的 1/4 实施、划分为 8 个区组:

定义方程：　　　$I=ABCDG=ABEFH=CDEFGH$；

区组混杂：　　　$ACF=BDFG=BCEH=ADEGH$，

　　　　　　　　$BDE=ACEG=ADFH=BCFGH$，

　　　　　　　　$ABCDEF=EFG=CDH=ABGH$，

　　　　　　　　$CDF=ABFG=ABCDEH=EGH$，

　　　　　　　　$AD=BCG=BDEF=ACEFGH$，

　　　　　　　　$BCEF=ADEFG=ACH=BDGH$，

　　　　　　　　$ABE=CDEG=FH=ABCDFGH$。

　　本设计和上一设计相比，多用了 CDF 一列作区组的进一步划分。注意，当 ACF、BDE 和 CDF 被区组混杂时，它们的全部广义交互作用也被区组混杂。同时，每当一个效应被区组混杂时，与该效应相混杂的效应也都被区组混杂。由于 8 个区组的划分，不可避免地要混杂个别二级效应，这里选择为 AD 和 FH。

④4,2 水平的转换：

　　把两个 2 水平因素的 4 个水平组合和 1 个因素的 4 个水平对应起来，就可以把这两个 2 水平因素合成一个 4 水平因素。注意，这样一来，原来的两个因素之间的交互作用不复存在（已包含到新因素的 4 个水平之中）。例如，在设计①中，把 E 和 F 合成一个 4 水平因素(EF)，这里应把(EF)看作一个字母，并不存在交互作用 EF［正因为如此，在 E 和 F 的 4 个水平组合被改成 4 个不同的水平之后，可将 EF 列（$ABCD$ 列）从正交表 $L_{32}(2^{31})$ 中除去］。于是设计①变为：

定义方程：　　　　　$I=ABCD(EF)$；

区组混杂：　　　$ACE=BDF$，　$CD=AB(EF)$，

　　　　　　　　$ADE=BCF$。

　　这是 $4×2^4$ 析因的 1/2 实施、划分为 4 个区组的一个设计方案，其中定义效应 $ABCD(EF)$ 应看作由 5 个字母组成，被区组混杂的效应 $AB(EF)$ 也应看作由 3 个字母组成。在这个设计方案中，仍然只有一个二级效应 CD 为区组效应所混杂。

　　在设计②和③中，也可以把某两个字母比方说 G 和 H 解释为一个 4 水平因素(GH)，从而把它们看成是 $4×2^6$ 析因的 1/4 实施方案。在划分 4 个区组的设计②中，全部一、二级效应都没有被区组混杂，但在划分 8 个区组的设计③中，则有 $E(GH)$、AD 和 FH 三个二级效应被区组混杂。比原来的设计多了一个混杂。

7.2 三水平情形

在三水平正交表中，利用一列作区组划分时，表中的水平组合即被划分为 3 个区组，利用二列作区组划分，则每个区组再被三等分而变为 $3^2=9$ 个区组。对于后一情形，不仅排有区组因素的二列被区组混杂，而且这二列的广义交互作用列也被区组混杂。

7.2.1 比较方案

在区组混杂设计中，仍然要尽量避免低级（一、二级）效应同区组混杂。例如下面的两个混杂设计：

① $\dfrac{X}{ABC}$, $\dfrac{Y}{AB^2C}$, $\dfrac{XY}{A^2C^2=AC}$, $\dfrac{XY^2}{A^3B^4C^3=B}$;

② ABC , AB^2 , $A^2C=AC^2$, $B^2C=BC^2$ 。

都不理想。设计①表示两个区组因素被排在 ABC 和 AB^2C 列，从这二列的交互作用看出，兼有一、二级效应被区组混杂。而设计②中，被混杂的效应只限于二级和二级以上，故可认为②比①好。

7.2.2 参考方案

仿照二水平情形，不难理解下面是一个值得参考的 3^5 析因的 1/3 实施方案：

定义方程： $I=ABCDE=(ABCDE)^2$;

混杂关系： $ABC^2=ABD^2E^2=CD^2E^2$,

$AB^2D=AC^2DE^2=BC^2E^2$,

$ACD^2=AB^2CE^2=BD^2E$,

$BCD=AB^2C^2D^2E=AE$ 。

实现方法：在正交表 $L_{81}(3^{40})$ 中，把 4 个因素 A,B,C,D 排在基本列，第 5 个因素 E 排在 $ABCD$ 列 [把 $ABCD$ 写成 $(ABCD)^2$ ，相当于同列中的水平 1 和 2 互易]，由此得定义方程。再利用 ABC^2 和 AB^2C 两列划分区组。由于 $ABC^2(X)$ 和 $AB^2D(Y)$ 被区组混杂，其广义交互作用 $ACD^2(XY)$ 和 $BCD(XY^2)$ 也被区组混杂。将这些被区组混杂的效应一一遍乘定义方程，即得上述混杂关系。

8. 多因素优选法

优选法的目的是要找出最优工艺条件（指最优水平组合）。仅安排一次正交试验未必能立即找到此最优条件，但经过正交设计的试验分析，往往能表明在下一步试验中,应如何调整参试因素以及如何变动它们的水平,以期较快地达到或接近最优。

8.1 一般步骤

优选的目的是要找到最优工艺条件,而不是要准确知道每一因素对试验所产生的效应,以及它在试验中与其他因素的交互作用。(当然不是说,明确这些因素效应和交互作用无助于寻找最优工艺条件)。因此,不一定要使用具备交互作用列的正交表,而只要表中的行数(试验次数)和列数(可容纳因素的个数)都比较合适就行,如表 $L_{12}(2^{11})$,$L_{18}(2\times 3^7)$ 等。即使选用有交互作用列的正交表如 $L_8(2^7)$,$L_9(3^4)$,$L_{27}(3^{13})$ 等,也不妨尽量多排因素,而不必过多地顾虑效应的混杂问题。

利用正交设计进行多因素优选,方法甚为灵活,但大致包括三个步骤。

第一步,选用试验次数较少但能容纳参试因素的一张正交表,如 $L_8(2^7)$,$L_{12}(2^{11})$,$L_{18}(2\times 3^7)$,$L_{27}(3^{13})$ 等,希望通过这样的所谓"大面积撒网"的正交设计,有可能碰上较好的试验点(指较好的水平组合)。

第二步,选出少数几个主要因素,再利用较小的正交表,如 $L_4(2^3)$,$L_9(3^4)$ 等,围绕已得到的好试验点作析因或部分析因试验,希望通过这样的所谓"小范围侦察",摸清楚试验结果的局部变化规律。

第三步,确定主要因素的有利水平变化方向,推进试验结果。必要时,重新挑选主要因素或围绕已出现的更好的试验点做试验,并回到第二步。

8.1.1 应用举例

为了提高某有机硅化合物的产率,考虑参试因素及其水平如表 18 所示,并按正交表 $L_8(2^7)$ 安排试验如表 19 所示。

表 18

水平	蒸馏瓶 A	反应温度 B	升温时间 C	搅拌速度 D	吡啶用量 E	控温时间 F	原料配比 G
0	乙	240~245 ℃	1.5	慢	15%	3 小时	1:2.6
1	甲	260~265 ℃	2	快	25%	4 小时	1:3.4

表 19

试验号	A	B	G AB	C	D AC	E BC	F ABC	产率/%
1	0	0	0	0	0	0	0	13.2
2	1	0	1	0	1	0	1	19.3
3	0	1	1	0	0	1	1	67.2
4	1	1	0	0	1	1	0	40.3
5	0	0	0	1	1	1	1	22.4
6	1	0	1	1	0	1	0	27.0
7	0	1	1	1	1	0	0	52.0
8	1	1	0	1	0	0	1	42.6
m_0	38.7	20.5	29.6	35	37.5	31.8	33.1	
m_1	32.3	50.5	41.4	36	33.5	39.2	37.9	
m_1-m_0	−6.4	30.0	11.8	1	−4	7.6	4.8	

表 19 中的 m_i 是对应于各列第 i 水平的平均产率。例如 A 列下的

$$m_0 = \frac{1}{4}(13.2+67.2+22.4+52.0)=38.7。$$

根据 m_1-m_0 的绝对值大小初步判断 B、G、E 为主要因素,并按 m_1-m_0 的正负号选出较好的水平组合为 $B_1G_1E_1$(下标表示水平),这恰好与产率最高的第 3 号试验相符①。于是决定下一步试验围绕第 3 号试验点对 B、G、E 做 2^3 析因的 1/2 实施(图 6)。B、G、E 的水平变化如表 20 所示,其余因素的水平与第 3 号试验同。试验结果见表 20。

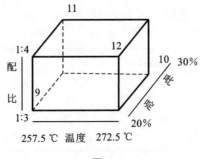

图 6

---

① 不相符的情形也是可能的,此时要对交互作用和试验误差进行适当的考察,方能判定较好的水平组合。

表 20

水平	反应温度 B	原料配比 G	吡啶用量 E
0	257.5 ℃	1:3	20%
1	272.5 ℃	1:4	30%

和前 8 次试验结果相比,最高产率从 67.2% 提高到 76.3%,说明主要因素的挑选是有效的。然而,从表 21 的 $m_1 - m_0$ 看,E 也许不再起主要因素作用。因此,在尔后推进产率的尝试中,只考虑变动 B 和 G 的水平。

表 21

试验号	B	G	E	产率/%
9	0	0	0	46.5
10	1	0	1	73.3
11	0	1	1	52.9
12	1	1	0	76.3
$m_1 - m_0$	25.1	4.7	1.7	

为了找出因素 B 和 G 的有利的水平变化方向,可采取的方法之一是所谓主效应比率法[①]。具体做法如下。

(1) 找出温度和配比变化的基点和单位。

温度的基点和单位:以温度的两个水平的平均值即 265 ℃ 为基点;以这两个水平之差的一半即 7.5 ℃ 为单位。

配比的基点和单位:以配比的两个水平的平均值即 3.5(表示1:3.5)为基点;以这两个水平之差的一半即 0.5 为单位。

(2) 计算两者变化的比例关系。

从基点起,按主效应比率 25.1:4.7(见表 21),温度每改变 25.1 个单位,配比相对地改变 4.7 个单位。

[①] 也可以配制所谓响应曲面,以估算一定范围内的产率。如

$$产率 = 62.25 + \frac{25.1}{2}B + \frac{4.7}{2}G + \frac{1.7}{2}E,$$

其中 62.25 为平均产率;B,C,E 的计算单位和原点(基点)如同后面所描述的。

根据本试验的化学反应的实际情况,温度不能超过 280 ℃。现取温度为 277.5 ℃,即温度改变

$$\frac{277.5-265}{7.5}=1.67 \text{ 个单位},$$

故配比要相应地改变

$$\frac{1.67\times4.7}{25.1}=0.313 \text{ 个单位},$$

即配比取为

$$3.5+0.313\times0.5=3.66\text{(表示 }1:3.66)\text{。}$$

实际试验结果为:

温度	配比	产率
277.5 ℃	1:3.66	81.3%

(次要因素吡啶为节约起见固定在 20%)。

8.1.2 主效应比率法的一般方法

确定温度与配比的变化比例关系之后,即定出一个优选方向,沿此方向选择的试验点称为调优点。其计算方法可程序化如下。

按主效应比率法计算调优点的程序

	温度	配比
基点	265 ℃	3.5(表示 1:3.5)
单位	7.5	0.5
效应	25.1	4.7
单位×效应	188.3	2.35
以温度 2.5 ℃为步长的相应水平改变值(注) 备用调优点:	2.5	$0.031\left(=\frac{2.5\times2.35}{188.3}\right)$
	272.5	3.59[①](有舍入误差)
	275.0	3.62
	277.5	3.66(实际调优点)
	280.0	3.69
	⋮	⋮

注:2.5 ℃仅是一个方便的数,也可取别的数,应视具体情况而定。

实际调优时,从计算出来的备用调优点中选做调优试验。在本例中,只选做了一点,一般可选若干点,看看产率是否继续得到提高。虽然从有

---

① 给定温度计算配比的公式:配比$=3.5+(\text{温度}-265)\times\dfrac{2.35}{188.3}$。

利的水平变化方向看,调优点离基点越远,产率会越高。然而,因素效应只代表试验范围的局部变化规律,调优点离基点太远,超过了有效的局部范围,就会得不到预期的效果。这时,如认为仍有潜力可挖,可另作正交设计,重新估计在变化了的水平范围的因素效应。

显然,主效应比率法不限于两个因素的情况。若认为需要考虑更多因素的效应,可对每个因素换算它的水平变化的基点和单位,然后以它们带有正负号的效应大小为比例,改变它们的水平。

8.2 映象设计

主要因素的筛选,首先要通过对主效应的分析和计算,而由于在部分析因尤其是高度的部分析因中,主效应同高级效应互相混杂,即使借助于专业知识,亦未必能正确判明因素的主次关系。在部分析因中,当因素的主效应大小尚不明朗时,常可加排一些试验,使之明朗化。特别对于二水平部分析因,主效应与二级效应之间的混杂,可通过所谓映象设计,使全部主效应不与二级效应相混杂。现举例说明。

假设在 $L_8(2^7)$ 表上排满了 7 个因素,做完 8 次试验之后,把表上的"0"水平和"1"水平全部对换,再做 8 次试验,那么后 8 次试验相对于前 8 次试验来说,就是一个映象设计。反过来,前 8 次试验相对于后 8 次试验来说,也是一个映象设计。事实上,可把上述 16 次试验分为两半,一半代表原象,另一半代表映象,叫哪一半为原象,哪一半为映象都可以。

若把映象添加到原象的下方,就是 $L_{16}(2^{15})$ 表上的最后 7 列,即第 9 列到第 15 列(见表 22)。容易验证,这后 7 列的任何二列的交互作用都落在该表的前 7 列上。例如,排在 AD 和 CD 二列上的两个因素的交互作用落在

$$AD \cdot CD = AC$$

列上。既然这些交互作用都不在最后的 7 列上,当映象与原象合并分析时,也就不发生主效应同二因素交互作用相混杂的问题。

表 22

试验号	二因素交互作用列								主效应列							过滤时间
	DM RS PQ NO	DN QS PR MO	DO QR PS MN	DP OS NR MQ	DQ OR NS MP	DR OQ NP MS	DS OP NQ MR		M	N	O	P	Q	R	S	
	1 A	2 AB B	3 AC C	4 ABC	5 AD BC	6 ABD D	7 ACD BD	8 ABCD CD	9 BCD	10	11	12	13	14	15	
								原象								
1	0	0	0	0	0	0	0	0	0	0	0	0	0	0	0	38.7
2	1	0	1	0	1	0	1	0	1	0	1	0	1	0	1	68.7
3	0	1	1	0	0	1	1	0	1	1	0	0	1	1	1	41.2
4	1	1	0	0	1	1	0	0	1	1	0	0	1	1	0	78.6
5	0	0	0	1	1	1	1	0	0	0	0	1	1	1	1	81.0
6	1	0	1	1	0	1	0	0	1	0	1	1	0	1	0	66.4
7	0	1	1	1	1	0	0	0	0	1	1	1	1	0	0	77.7
8	1	1	0	1	0	0	1	0	1	1	0	0	0	0	1	68.4
								映象								
9	0	0	0	0	0	0	0	1	1	1	1	1	1	1	1	67.6
10	1	0	1	0	1	0	1	1	0	1	0	1	0	1	0	42.6
11	0	1	1	0	0	1	1	1	0	0	1	1	0	0	0	59.0
12	1	1	0	0	1	1	0	1	0	0	1	1	0	0	1	47.8
13	0	0	0	1	1	1	1	1	1	1	1	0	0	0	0	61.9
14	1	0	1	1	0	1	0	1	0	1	0	0	1	0	1	86.4
15	0	1	1	1	1	0	0	1	1	0	0	0	0	1	1	65.0
16	1	1	0	1	0	0	1	1	0	0	1	0	1	1	0	66.7
m_1-m_0	4.2	−1.1	−0.5	15.2	3.6	3.4	−4.9		6.7	3.9	−2.7	0.4	19.2	1.0	4.3	

注：这里 m_i 的意义与表 20 中相同。

事实上还可以利用第 8 列（D 列），安排第 8 个因素，并且这 8 个因素中的任何两个因素的交互作用都不至于同它们的主效应相混杂。因此,若把前后两批 8 次试验看作两个区组,也不致发生区组与主效应相混杂的问题。

映象设计并不限于对 $L_8(2^7)$ 映象,还可应用于其他的 2 水平正交表如 $L_4(2^3)$, $L_{12}(2^{11})$ 等表上。

8.2.1 应用举例

为了缩短某工艺中的过滤时间,以加快生产过程,考虑 7 个可能影响过滤时间的因素。按正交表 $L_8(2^7)$ 安排如下试验(表 23,表 24):

表 23 因素与水平

水平	因素						
	M 水源	N 原料	O 再循环	P 过滤温度	Q 苛性钠输入速度	R 过滤布	S 存放搅拌槽时间
0	井水	外地	无	高	慢	长	原有
1	自来水	本地	有	低	快	短	新

表 24 试验结果的分析与计算

试验号	M	N	O	P	Q	R	S	过滤时间
				$L_8(2^7)$				
1	0	0	0	0	0	0	0	38.7
2	1	0	1	0	1	0	1	68.7
3	0	1	1	0	0	1	1	41.2
4	1	1	0	0	1	1	0	78.6
5	0	0	1	1	1	1	0	81.0
6	1	0	1	1	0	1	0	66.4
7	0	1	1	1	1	0	0	77.7
8	1	1	0	1	0	0	1	68.4
m_0	59.7	63.7	66.7	56.8	53.7	63.4	65.4	
m_1	70.5	66.5	63.5	73.4	76.5	66.8	64.8	
m_1-m_0	10.8	2.8	−3.2	16.6	22.8	3.4	−0.6	

计算表明,对应于 M、P、Q 三列的 m_1-m_0 的绝对值较大;由交互作用列的推导得知,这三列的任一列都是其余二列的交互作用列。通常认为,主效应大的因素其交互作用也很可能大。因此,对以上试验结果,可作如下四种可能解释:

(1) M、P、Q 的主效应均较大,故认为 M、P、Q 都是主要因素;

(2) M、P 的主效应和 MP 交互作用较大,故认为只有 M、P 两个主要因素;

(3) M、Q 的主效应和 MQ 交互作用较大,故认为只有 M、Q 两个主要因素;

(4) P、Q 的主效应和 PQ 交互作用较大,故认为只有 P、Q 两个主要因素。

但毕竟还不知道哪一种解释正确。为此,补充一个映象设计如表 25 所示。

表 25 映象设计

试验号	M	N	O	P	Q	R	S	过滤时间
9	1	1	1	1	1	1	1	67.6
10	0	1	0	1	0	1	0	42.6
11	1	0	0	1	1	0	0	59.0
12	0	0	1	1	0	0	1	47.8
13	1	1	1	0	0	0	0	61.9
14	0	1	0	0	1	0	1	86.4
15	1	0	0	0	0	1	1	65.0
16	0	0	1	0	1	1	0	66.7
$m_1 - m_0$	2.5	5	−2.3	−15.8	15.6	−3.3	9.2	

这后 8 次映象试验给出的 $m_1 - m_0$ 值和前 8 次原象试验相比,颇有出入,特别是 P 列的前后两个 $m_1 - m_0$ 值一负一正,相去甚远。足见只根据原象试验进行主效应分析未必能作出正确结论,需要把原象和映象合起来一并分析(见表 22)。计算结果表明,M、Q 两个主效应较大,而且交互作用较大的列也正是 MQ 所在的列(虽然交互作用 MQ、NR 和 OS 混同在一列上,但通常倾向于认为,主效应较大的因素,其交互作用也会较大,故认为该列主要代表 MQ),从而断定 M、Q 为主要因素。

按过滤时间越短越好的要求,应选取水平组合 M_0Q_0,即采用井水和慢速输入苛性钠。继续试验表明,这一措施使过滤时间稳定地下降,收到良好效果。

8.2.2 交互作用的分析

通常,若发现正交表中某交互作用列有较大的极差(较大是相对于所估计的试验误差而言),则需要对有关因素选择其较优水平搭配。从表22看到,第4列有较大的极差,而该列代表交互作用 MQ、NR、OS 和 DP,因此,如果认为这些交互作用都可能是真实的,就应找出 M 和 Q,N 和 R 等每一对因素的较优水平搭配。如上分析,MQ 被认为是真实的,于是将 M 和 Q 的不同水平组合所对应的平均试验结果列表(见表26)进行比较,从而明显看到 $M_0 Q_0$ 为较好的水平搭配。这和按主效应分析的结果一致(如不一致,便要继续做试验判明究竟什么水平搭配为好)。同样,若认为必要,也可以对 NR 和 OS 进行类似的分析计算。至于 DP,若认为区组与过滤温度交互作用的可能性不大,就不必去管它了。

表 26

	M_0	M_1
Q_0	$\frac{1}{4}(38.7+41.2+42.6+47.8)=42.6$	$\frac{1}{4}(66.4+68.4+61.9+65.0)=65.4$
Q_1	$\frac{1}{4}(81.0+77.7+86.4+66.7)=78.0$	$\frac{1}{4}(68.7+78.6+67.6+59.0)=68.5$

9. 几种常用正交表

9.1 两个到多个因素、各二个水平的 $L_{2^i}(2^k)$ 型正交表

$L_{16}(2^{15})$
$L_8(2^7)$
$L_4(2^3)$

试验号	A	B	AB	C	AC	BC	ABC	D	AD	BD	ABD	CD	ACD	BCD	ABCD
1	0	0	0	0	0	0	0	0	0	0	0	0	0	0	0
2	1	0	1	0	1	0	1	0	1	0	1	0	1	0	1
3	0	1	1	0	0	1	1	0	0	1	1	0	0	1	1
4	1	1	0	0	1	1	0	0	1	1	0	0	1	1	0
5	0	0	0	1	1	1	1	0	0	0	0	1	1	1	1
6	1	0	1	1	0	1	0	0	1	0	1	1	0	1	0
7	0	1	1	1	1	0	0	0	0	1	1	1	1	0	0
8	1	1	0	1	0	0	1	0	1	1	0	1	0	0	1
9	0	0	0	0	0	0	0	1	1	1	1	1	1	1	1
10	1	0	1	0	1	0	1	1	0	1	0	1	0	1	0

续表

试验号	A	B	AB	C	AC	BC	ABC	D	AD	BD	ABD	CD	ACD	BCD	ABCD
11	0	1	1	0	0	1	1	1	1	0	0	1	1	0	0
12	1	1	0	0	1	1	0	1	0	0	1	1	0	0	1
13	0	0	0	1	1	1	1	1	1	1	0	0	0	0	0
14	1	0	1	1	0	1	0	1	0	1	0	0	1	0	1
15	0	1	1	1	1	0	0	1	1	0	0	0	0	1	1
16	1	1	0	1	0	0	1	1	0	0	1	0	1	1	0

注：为方便使用起见，可将表中号码 0、1，通过对应关系 0→1，1→2，分别换成 1、2。

9.2 两个到多个因素、各三个水平的 $L_{3^k}(3^k)$ 型正交表

$L_{27}(3^{13})$

试验号	$L_9(3^4)$												
	A	B	AB	AB^2	C	AC	AC^2	BC	BC^2	ABC	ABC^2	AB^2C	AB^2C^2
1	0	0	0	0	0	0	0	0	0	0	0	0	0
2	1	0	1	1	0	1	1	0	0	1	1	1	1
3	2	0	2	2	0	2	2	0	0	2	2	2	2
4	0	1	1	2	0	0	0	1	1	1	1	2	2
5	1	1	2	0	0	1	1	1	2	2	2	0	0
6	2	1	0	1	0	2	2	1	0	0	0	1	1
7	0	2	2	1	0	0	0	2	2	2	2	1	1
8	1	2	0	2	0	1	1	2	0	0	0	2	2
9	2	2	1	0	0	2	2	2	1	1	1	0	0
10	0	0	0	0	1	1	1	1	2	1	2	1	2
11	1	0	1	1	1	2	0	1	2	2	0	2	0
12	2	0	2	2	1	0	1	1	2	0	1	0	1
13	0	1	1	2	1	1	2	2	0	2	0	0	1
14	1	1	2	0	1	2	0	2	0	0	1	1	2
15	2	1	0	1	1	0	1	2	1	1	2	2	0
16	0	2	2	1	1	1	2	0	1	0	1	2	0
17	1	2	0	2	1	2	0	0	1	1	2	0	1
18	2	2	1	0	1	0	1	0	1	2	0	1	2

续表

试验号	A	B	AB	AB^2	C	AC	AC^2	BC	BC^2	ABC	ABC^2	AB^2C	AB^2C^2
19	0	0	0	0	2	2	1	2	1	2	1	2	1
20	1	0	1	1	2	0	2	2	1	0	2	0	2
21	2	0	2	2	2	1	0	2	1	1	0	1	0
22	0	1	1	2	2	2	1	0	2	0	2	1	0
23	1	1	2	0	2	0	2	0	2	1	0	2	1
24	2	1	0	1	2	1	0	0	2	2	1	0	2
25	0	2	2	1	2	2	1	1	0	1	0	0	2
26	1	2	0	2	2	0	2	1	0	2	1	1	0
27	2	2	1	0	2	1	0	1	0	0	2	2	1

注：为方便使用起见，可将表中号码0、1、2，通过对应关系 0→1,1→2,2→3，分别换成1、2、3。

9.3 二、三水平混合正交表

$L_{18}(2\times 3^7)$

试验号	列号							
	1	2	3	4	5	6	7	8
1	1	1	1	1	1	1	1	1
2	1	1	2	2	2	2	2	2
3	1	1	3	3	3	3	3	3
4	1	2	1	1	2	2	3	3
5	1	2	2	2	3	3	1	1
6	1	2	3	3	1	1	2	2
7	1	3	1	2	1	3	2	3
8	1	3	2	3	2	1	3	1
9	1	3	3	1	3	2	1	2
10	2	1	1	3	3	2	2	1
11	2	1	2	1	1	3	3	2
12	2	1	3	2	2	1	1	3
13	2	2	1	2	3	1	3	2
14	2	2	2	3	1	2	1	3
15	2	2	3	1	2	3	2	1

续表

试验号	列号							
	1	2	3	4	5	6	7	8
16	2	3	1	3	2	3	1	2
17	2	3	2	1	3	1	2	3
18	2	3	3	2	1	2	3	1

注：(1)本表无交互作用列，但可通过通常的两向分类方差分析方法考察 1、2 两列的交互作用。

(2)将表中 1、2 两列的水平搭配"1,1"，"1,2"，…，"2,3"依次换成"1"，"2"，…，"6"，就可得到一张 $L_{18}(6 \times 3^6)$ 表。

9.4 由原象和映象构成的 $L_{24}(12 \times 2^{12})$ 混合水平表

$L_{24}(12 \times 2^{12})$

试验号	$L_{12}(2^{11})$											12	13
	1	2	3	4	5	6	7	8	9	10	11		
	原象												
1	1	1	1	1	1	1	1	1	1	1	1	1	1
2	1	1	1	1	1	2	2	2	2	2	2	1	2
3	1	1	2	2	2	1	1	1	2	2	2	1	3
4	1	2	1	2	2	1	2	2	1	1	2	1	4
5	1	2	2	1	2	2	1	2	1	2	1	1	5
6	1	2	2	2	1	2	2	1	2	1	1	1	6
7	2	1	2	2	1	1	2	2	1	2	1	1	7
8	2	1	2	1	2	2	2	1	1	1	2	1	8
9	2	1	1	2	2	2	1	2	2	1	1	1	9
10	2	2	2	1	1	1	1	2	2	1	2	1	10
11	2	2	1	2	1	2	1	1	1	2	2	1	11
12	2	2	1	1	2	1	2	1	2	2	1	1	12
	映象												
13	2	2	2	2	2	2	2	2	2	2	2	2	1
14	2	2	2	2	2	1	1	1	1	1	1	2	2
15	2	2	1	1	1	2	2	2	1	1	1	2	3
16	2	1	2	1	1	2	1	1	2	2	1	2	4

续表

试验号	1	2	3	4	5	6	7	8	9	10	11	12	13
17	2	1	1	2	1	1	2	1	2	1	2	2	5
18	2	1	1	1	2	1	1	2	1	2	2	2	6
19	1	2	1	1	2	2	1	1	2	1	2	2	7
20	1	2	1	2	1	1	1	2	2	2	1	2	8
21	1	2	2	1	1	1	2	1	2	2	2	2	9
22	1	1	1	2	2	2	1	2	1	2	1	2	10
23	1	1	2	1	2	1	2	2	2	1	1	2	11
24	1	1	2	2	1	2	2	1	1	2	1	2	12

注:(1)原象和映象的关系是:将原象部分的水平号码"1"和"2"对调,就得到映象部分。因此映象本身也是一张 $L_{12}(2^{11})$ 正交表。

(2)若只利用 $L_{24}(12\times 2^{12})$ 的前12列安排试验因素,则在高级交互作用(指三个或多于三个因素之间的交互作用)可忽略的前提下,因素的主效应将不与任何二因素之间的交互作用相混杂。

(3)1—11列是12与13两列的交互作用列。

9.5 四水平正交表

$L_{16}(4^5)$

试验号	列号				
	1	2	3	4	5
1	1	1	1	1	1
2	1	2	2	2	2
3	1	3	3	3	3
4	1	4	4	4	4
5	2	1	2	3	4
6	2	2	1	4	3
7	2	3	4	1	2
8	2	4	3	2	1
9	3	1	3	4	2
10	3	2	4	3	1
11	3	3	1	2	4
12	3	4	2	1	3

续表

试验号	列号				
	1	2	3	4	5
13	4	1	4	2	3
14	4	2	3	1	4
15	4	3	2	4	1
16	4	4	1	3	2

注(1):表中任二列的交互作用列是另外三列。
(2):表中任一个四水平列都可改为三个二水平列。例如,为了把第五列改为三个二水平列,可取如下的对应关系:

  1→1 1 2
  2→2 1 1
  3→1 2 1
  4→2 2 2

即将第五列的 1 变为三个数 1,1,2,这三个数各占一列;将第五列的 2 变为三个数 2,1,1,这三个数也依次地各占一列;等等。这样,第五列就变为右边的三列,从而得到一张 $L_{16}(4^3×2^3)$ 表。这个办法反复地施用于其他每一列上,就会得到 $L_{16}(4^3×2^6)$, $L_{16}(4^2×2^9)$, $L_{16}(4×2^{12})$ 和 $L_{16}(2^{15})$ 等正交表。

右侧对应列(三个二水平列):

1 1 2
2 1 1
1 2 1
2 2 2
2 2 2
1 2 1
2 1 1
1 1 2
2 1 1
1 1 2
2 2 2
1 2 1
1 2 1
2 2 2
1 1 2
2 1 1

9.6 五水平正交表

$$L_{25}(5^6)$$

试验号	列号					
	1	2	3	4	5	6
1	1	1	1	1	1	1
2	1	2	2	2	2	2
3	1	3	3	3	3	3
4	1	4	4	4	4	4
5	1	5	5	5	5	5
6	2	1	2	3	4	5
7	2	2	3	4	5	1
8	2	3	4	5	1	2
9	2	4	5	1	2	3
10	2	5	1	2	3	4
11	3	1	3	5	2	4

续表

试验号	列号					
	1	2	3	4	5	6
12	3	2	4	1	3	5
13	3	3	5	2	4	1
14	3	4	1	3	5	2
15	3	5	2	4	1	3
16	4	1	4	2	5	3
17	4	2	5	3	1	4
18	4	3	1	4	2	5
19	4	4	2	5	3	1
20	4	5	3	1	4	2
21	5	1	5	4	3	2
22	5	2	1	5	4	3
23	5	3	2	1	5	4
24	5	4	3	2	1	5
25	5	5	4	3	2	1

注：表中任二列的交互作用列是另外四列。

10. "正交设计法"相关和反馈信息

10.1 读者来信摘要

[编者按]"正交设计在农业试验上的应用(1—9讲)"是20世纪70年代林少宫教授写在祖国大地上的论文。该论文诠释部分析因设计的原理与应用，涵盖作者多次深入工厂、农村、田野为正交试验设计法所做的调研和推广及所取得的成果，真正做到理论联系实际。当时，文章在《湖北农业科学》科技讲座专栏发表后，给农业生产试验带来了积极的指导作用，也产生了广泛的学术影响，在全国范围内深受基层科研人员的欢迎，与此同时，也出现刊载该讲座的《湖北农业科学》供不应求的情况，以至不少读者写信向作者求助。

四川省潼南县农科所付大雄
1979年10月23日

敬爱的林老师:读了您在《湖北农业科学》上的"正交设计在农业试验上的应用"的讲座,对我们深入研究正交设计法帮助极大。国内公开发行的有关正交设计的书我们都读过,您的讲座独具一格,深入浅出,令人感觉到您广博的知识、雄厚的数理统计基础。但是我们费了好大的气力都找不齐讲座的全文,目前还差1、2、3、8讲没有读到,我们多么希望读完全部讲座啊!

福建省永太县北斗农场郭坤池
1979年1月20日

林少宫老师:您在《湖北农业科学》1977年11期至1978年12期发表的关于"正交设计在农业试验上的应用",我们学习了数遍,认为写得很好,对提高试验的准确性和我们的分析能力很有帮助,因此,我们建议您将上述的文章印成单行本,供广大农业科技人员学习、使用。

云南省思茅地区农业科学研究所
1979年3月19日

华中工学院:贵院林少宫同志编讲的"正交设计在农业试验上的应用"在《湖北农业科学》1977年10~12期,1978年1~11期连载,这对我们搞农业科研的帮助大,我们这里买不到这几期杂志,若贵院印有讲义,请支援35~40本,工本费由我所汇还。

安徽省白湖农场农科所印天寿
1979年2月25日

林少宫老师:您在《湖北农业科学》上连篇刊载的"正交设计在农业试验上的应用"九讲,我一一认真地读了,深感用处很大,确实对农业科学试验研究工作者的帮助不小。我一遍又一遍地读着,已读过两遍了,现在开始读第三遍,感到内容是比较深的,特别是最后几讲领会不透之处较多(由于我的水平差)。因此,十分迫切地希望您能够赐予帮助,不知是否有较为详尽的有关讲义?

四川省广汉县示范繁殖农场邓启明
1977年11月10日

少宫同志:您好!阅《湖北农业科学》1977年第10期刊载您写的"正交

设计在农业试验上的应用"一文之后,我们认为写得很好,很适合四级农科网的同志们学习和采用,这是您对农业学大寨和科学种田做出的一项新贡献,对此,我表示敬佩!

您写的这篇文章的全文,我们想翻印成册,供四级农科网的同志们学习,特此联系。希能把这篇文章的全文的复制稿寄一份给我们。

浙江省余杭县长岗农场生产组
1978 年 6 月 10 日

林少官老师:我们在《湖北农业科学》杂志上看到您所写的"正交设计在农业试验上的应用"一文,感到很有用,很有启示,望能把全文寄给我们。

黑龙江省安庆县农科所王世录
1979 年 3 月 5 日

亲爱的林老师:您好!我是安庆县人,北安农校五七年中专毕业,一直从事农业科研工作,二十几年一直是在学习,可是从来也未有学您的著作如此解渴。1978 年 6 月,我见到您写的"正交设计在农业试验上的应用"一文,很感兴趣,所以我就千方百计地买《湖北农业科学》杂志,但是一直到今天还是没有集齐,还缺少第一、四、八讲,只好硬着头皮写信,向老师您求援。

广东省新会县科学技术委员会叶耀南
1978 年 9 月 29 日

林少官教授:去年有一面之缘,蒙教诲,深得益。如 $L_8(2^7)$ 的运用和田间试验设计等。尔后,在实际工作中做了应用的尝试。在此,深表谢意!

最近,在《湖北农业科学》中见到您的大作,拜读之余,也做了四讲笔记抄录,从已发表的七讲中仅七分之四。但限于时间未能全部抄完。在抄录中思绪起伏,每念及能收录全部印刷材料的话则省掉总是抄写的时间,能多看些资料。多次想写信给您索取该资料,欲想写而又停。近两天来考虑再三,毅然执笔。

湖北省沔阳县通海公社东方红大队王圣荣
1979 年 10 月 11 日

尊敬的林教授:您好!您邮来的资料本人于十月五日收到。阅后,我感到非常高兴。这些资料对于我学习科学理论进行科学试验都有着重要的指导意义。这也是您对我们农业基层技术员的最大关怀和鼓励。在此，我谨向您表示由衷的谢意。

10.2 访谈摘要

这次访谈，林老师花了较多时间，是围绕第二桩事"正交试验设计"的话题展开的。从中我们听得出，林老师对当年"正交试验设计"那段工作经历感受很深，是颇为留念、珍惜的。同时我们更感觉到，林老师是在据此而作出引伸，使之在更为广泛意义上，作出我们现场统计研究工作的回顾与展望，以及启发我们思考，如何在不同环境条件下去开拓、创新我们的现场统计研究工作。

关于正交设计的研究和应用，好像昨天发生的事情那样，林老师记得清清楚楚，谈了许多"现场"故事和思想。他说：

"试验设计是数理统计的一个专门的领域，析因设计及其部分实施又是试验设计中的极具应用性而比较抽象的一类理论成果。"

"我们所研究推广的正交试验，经过推广应用一段时间之后需要有所提高，以做到更好的深入浅出。浅出才能推广，深入才足以引起试验工作者的持久兴趣。为此我写了一系列的论文，有我们创新的东西。"

回顾在 1972—1978 年的一段较长的时间里，林老师和华工（现华中科技大学）数学系的一些年轻教师跑了许多工厂、农村、田野，我们感到真正做到了现场统计的研究。

林老师于是说：

"现场能给我们很多启发，使我们及时去钻研很多有价值的问题，去解决现场的问题。"

"把试验设计用到现场，就是去发现它的价值，实现它的价值。"

"现场有很多我们不懂的东西，我们是从不懂到懂的；现场遇到的很多实际问题，使人很受启发，创新就是在这样的过程中形成的。"

林老师继续说：

"在学术上有不同的风格、不同的观点，这样才好；有学术探讨，才有可能开拓、创新。对某种格式或方法仅停留于相互抄袭，是无意义的。"

"某一国外专家，在正交试验设计方面确乎有较高的知名度，但据说他不太注重数学，在处理某些理论与实际应用问题时，数学上糊糊涂涂，是不可取的。"

"把部分析因与现代应用上的正交设计,从理论上联系起来对应用很有实际价值。"

"我们当时是从实际问题的研究来学习一些有关的理论知识的,研究需要具备一定的理论基础,才有可能做到开拓、创新。"

在谈到我们研究会工作问题时,林老师颇有感慨地说:

"我们当时的处境,远远不能和现在这么好的环境条件相比,当时既无研究编制又无经费来源,也许是想做些真正有用、有益的事驱使我们有干劲、有灵感、有成效,而不计报酬、名利。我现在仍十分怀念那段工作经历,可惜没让我一直做下去,因为后来要我去搞经济了。"

在林老师谈到有关数量经济教学与研究的话题时,颇令我们为之惊赞的是,他居然把"试验设计"的思想、方法融合于经济学中。

来源:湖北省现场统计研究会吕梓琴:《专访林少宫教授》2001年5月

笔者曾采访过一些林少宫教授带过的学生和他的同事。华中科技大学数学系李楚霖教授说:"林老师一直是数学系以及经济学院老师的榜样,是中国统计学界难得的科学家。他从年轻时起,就一直热衷于统计学研究,从未间断,并且时常有自己的创新。这些成果的取得不是偶然的,除了他的天赋和勤奋之外,他那种不计较名利、不惧怕艰辛的精神,正是我们很多年轻人所缺少的。他对这个学科在国内的发展起到至关重要的作用。"

他在正交试验设计领域的研究产生了巨大的影响,有广泛而重要的应用价值,推动了部分析因试验设计在我国工农业生产实验的因素分析中的广泛应用,创造巨大的经济效益和社会效益。

华中科技大学经济学院朱松青:《林少宫教授的治学精神》2001年6月

10.3 文章摘要

正交设计的推广应用,已在全国范围内获得巨大的成功。这是我们实行了三结合的结果。第一,组织领导重视,有组织有计划地抓推广应用工作。第二,走群众路线,紧密结合群众的生产经验和专业知识。第三,数学工作者走与工农相结合的道路,数学方法一再简化、表格化,使正交试验设计做到"简单、有效",适应广大工农群众的需要。任何试验项目,要应用正交设计取得成功,都要贯彻执行三结合的方针。

林少宫:《正交试验设计讲座(农机试验例解)》1976年

实(试)验设计是数理统计的一个专门领域,析因设计及其部分实施又是实验设计中的极具应用性而比较抽象的一类理论成果。

我们所研究推广的正交试验,本来自农业试验的析因设计,经过了一段时间在工业生产方面的推广应用后,再返回到农业是很自然的,但这时需要有所提高,深入浅出。浅出才能推广,深入才足以引起试验工作者的持久兴趣。为此,我写了《正交设计在农业试验上的应用》系列论文;为了简化分析计算,我和吕梓琴研制成《极值分析临界值表》,甚受欢迎。在工业试验方面,我从因素筛选,到定向探索,再到优化推进,拟定了一套简单易懂而有普遍应用意义的模式。这一模式后来竟被引用到正交试验的教材中。

在这几年里,我和数学系的一些年轻教师跑了许多工厂、农村、田野,真正做到了现场统计研究。

林少宫:《简短的回顾》2001年5月

1972年以后,林少宫老师带领试验设计组先后在武汉粉末冶金厂、武汉化工研究所等单位,结合现场问题研究项目,运用正交试验方法,取得成效。1974年暑期,在市政府招待所,林老师一行为武汉地区有关工矿企业、科研部门的科技人员举办正交试验设计方法应用讲习班。1975年2、3月,林老师一行对广东省在农业上应用正交试验设计法,进行了为期月余的考察,考察了花县、番禺、肇庆、新会、台山、阳江、信宜、高州等地市县。从广东考察返汉后,林老师团队立即把正交试验法从工业推展到农业的普及应用上,先后多次去湖北潜江、沙市、荆州、沔阳、宜昌、孝感、通城、黄州等地,进行了有声有色、卓有成效的普及和推广应用活动,在多个项目中取得良好效果,如沔阳沙湖原种场的"早稻品比及栽培试验"、联合公社农技站的"早稻栽培试验"等。1977年3月中、下旬,林少宫老师应广东省科技局之邀,赴广州、江门、新会等地讲学咨询,吕梓琴老师随同前往;返汉途中,又一同前往湖南省农科院(袁隆平所在处),临别前,他们获赠一面锦旗,上面写着"传经送宝情意重,工为农想风格高"字样。

(根据《湖北省现场统计研究会通讯》第5期/2002.12综合整理)

10.4 看法与建议

关于目前在农业科学试验中应用"正交法"及其他数学方法的建议

执笔:林少宫(1977年4月)

为加速我国社会主义大农业朝着现代化的目标发展,综合利用各种近代先进科学技术成果,有关数学方法如概率统计、运筹学等是不可缺少的重要内容和有力工具之一。然而数学方法在农业上的应用,相对于其他学科而言,是一个薄弱环节。近年来,统筹法、优选法在工农业上的应用比较成熟,而其他数学方法包括正交法在内,在农业上的应用仅仅是有了一个良好开端,还存在一些问题。为此,我们提出一些初步看法与建议,呈请有关领导部门给予重视。

(1) 目前,我国农业试验设计的学科内容比较陈旧,离现代化的要求较远,不利于我国农业科学研究和生产的发展,需要组织农学和数学工作者通力合作,对这个学科内容进行改造与创新工作,这是一项摆在我们面前艰巨而有意义的任务。

(2) 近年来,农业应用"正交法"先后在广东、湖北两省逐步开展,已有良好的群众实践基础,它对于解决多因素的试验方法问题,起了重要作用。实践证明,正交法在饲养学、组织离体学、栽培学以及微生物学等方面应用,效果很好,能迅速摸清规律并及时指导大生产,是农业试验方法的一种创新。但也有一些项目,由于方法使用不当,而导致不可靠的结论,甚至引出错误论断。因此,必须组织有关人员对这种方法作进一步的研究和总结提高,把正交法在农业上的应用,提高到一个新的水平。

(3) 我们认为进一步研究的主要内容是:正交法在农业哪些范围运用?如何系统地整理出一套试验内容的设计与田间安排以及相应的统计分析方法。要完成此任务,须做大量工作,不仅要认真、深入总结我们两省的实践经验,并虚心学习兄弟省市的好经验,还要收集、整理国内外有关科技资料,不仅涉及数学方法问题,还要涉及农业科学知识。这就需要有组织,有计划,分工协作地进行工作。建议两省首先通过数学等专业学会形式,建立这种方法应用研究协作关系,由两省科技、教育有关部门负责组织、协调工作。

[后记] 根据 1975 年 10 月 15 日中国科学院文件(75)科发二字 716 号《概率统计会议纪要》建议之一:"请湖北、广东两省科技局协商召开农业应用概率统计会议,就各项任务的分工协作,作出具体安排。"当时,湖北、广东两省在农业生产和科学试验中应用正交设计方面都做过一些工作。1976 年 4 月,由林少官老师牵头的华中工学院试验设计小组,在湖北省科委的领导和支持下,同广东省科技局推优办,就农业应用概率统计,特别是正交设计,以及其他数学方法的问题,充分交换了意见,提出一些共同看法和建议,建议书由林老师起草。

第四部分
数量经济

  1980年夏,诺贝尔经济学奖获得者克莱因(Lawrence Klein)教授率领美国经济学家代表团与中国社会科学院合作,在北京颐和园举办了为期7周的"计量经济学讲习班",这成为中国数量经济学发展历程中的一个标志性事件,林少官教授被特邀担任此次讲习班的翻译并讲解,为讲习班的成功举办起到了重要作用。随后,林教授率先在华中工学院(现华中科技大学)开始数量经济学的教学与研究工作。1982年4月,华中工学院数量经济研究所正式成立,林教授出任首任所长。该学科创建初期,除了应对客观条件的困难,林教授还考虑到了如何进行高水平和系统的经济学教育,以及如何迅速提高学生的经济研究能力。林教授借鉴国际上经济学教育理念,并结合国内实际情况,制定了切实可行的培养方案。对一些主干课程,林教授身体力行,坚持用英语授课,教材和教学内容尽可能和国际一流学校接轨。他上课大多采用国际通用教材或讲义,其中有些是从美国带回来的,有些是国际著名学者讲学时赠送的。1983年,林教授邀请麻省理工学院经济学教授费歇尔(F. M. Fisher)来校讲授微观经济学,之后,推荐其微观经济学讲义在中国翻译出版。1984—1985年,林教授推荐其1982届研究生参与国家教委和普林斯顿大学邹至庄教授主办的经济学福特培训班,并应国家教委之约,和中国人民大学高鸿业教授一起为该班学员的选拔和录取出考题。平时,林教授重视学以致用,给学生安排实践课题,如中国社会科学院投入—产出的课题实践、湖北省经济委员会经济增长的课题计算等。这些社会工作经验,不仅有利于了解中国的国民经济运行情况,对日后从事经济学研究也有潜移默化的作用。2000年,年近八旬的林教授不辞辛劳地翻译完古扎拉蒂的《计量经济学》上、下册(约120万字),之后,还主持翻译了伍德里奇的《计量经济学导论》。为引进计量经济学和培养现代经济学人才,林教授做了大量奠基性的工作。

  林教授在数量经济学方面的著作包括《简明经济统计与计量经济》《微观计量经济学要义》等;论文包括"微观计量经济学中的实验设计法""计量经济与计量金融"等。部分论文被本章收录。

正交实验设计在微观计量经济中的应用
——政策（或事件）评价法[①]

编者按：本期《通讯》我们发表了两篇尚未公开发表的报告与论文，一是中国科学院院士陈希孺研究员 2001 年 12 月在华中科技大学数学系作的题为《统计大样本理论杂谈》演讲稿，二是华中科技大学经济学院林少宫教授、博导的近作《正交实验设计在微观计量经济中的应用——政策（或事件）评价法》，以飨读者朋友们。

同时，我们还登载了《湖北省现场统计研究会第四次代表大会概况》，以向因故未能出席大会的会员与同行朋友通报大会情况、研究会动态以及第三届理事会简要工作总结。

内容提要：人们早已意识到政策（事件）分析在计量经济学的应用中占有重要份量，却很少知道正交实验设计能在其中起什么作用。本文借助 Wooldridge 所引用的 Kiel 和 McClain 一例，评价在某城市兴建一座垃圾焚化炉对当地房价的影响，对比了回归分析和 $L_4(2^3)$ 分析的结果，以阐明正交设计的用途及思想方法。

关键词：交互作用　回归设计　正交设计　综列数据

1. 回归设计与分析

1978 年开始传言要在某市兴建一座垃圾焚化炉，1981 年动工了，1985 年投入运转。

---

① 原文发表于《湖北省现场统计研究会通讯》2002 年第 4 期。

天真的分析者仅仅使用1981年的数据,估计了一个非常简单的模型:
$$P = r_0 + r_1 N + u \tag{1}$$
其中,P代表住房的平均真实价格(或不变价格),N代表虚拟变量(或二分变量):
$$N = \begin{cases} 1 & \text{如果住房靠近焚化炉(3公里以内)} \\ 0 & \text{如果住房远离焚化炉} \end{cases}$$
u代表通常的误差项。估计的方程为:

1981年:
$$\hat{P} = \underset{(3093.0)}{101307.5} - \underset{(5827.7)}{30688.27} N \qquad n = 142, \qquad R^2 = 0.165 \tag{2}$$

括号内为估计量的标准误,n为样本大小,而R为相关系数。估计结果表明,靠近焚化炉的住房平均售价比远离焚化炉的要低\$30688.27,并且这个差别是统计上显著的。但能说焚化炉的座落是造成附近房价平均下跌约\$30.7(千)的(全部)原因吗?

现在我们再利用1978年(尚无焚化炉传言之前)的数据做同样的回归,又得到

1978年:
$$\hat{P} = \underset{(265379)}{82517.23} - \underset{(5827.71)}{18824.37} N \qquad n = 179, \qquad R^2 = 0.082 \tag{3}$$

表明原来靠近焚化炉座落的平均房价就比远离它的要低\$18824.37,而且这也是统计上显著的,从而支持了焚化炉本来就要建造在房价较低地带的观点。也就是说,低房价地带是因,焚化炉的选址是果,而不是反过来。

怎样分析其中的因果关系呢?关键在于理解两个不同"年份"和距离焚化炉的"远近"之间的交互作用。这个交互作用就由
$$\delta = (\hat{P}_{81,N} - \hat{P}_{81,F}) - (\hat{P}_{78,N} - \hat{P}_{78,F}) = -30688.27 - (-18824.37)$$
$$= -11863.9$$

来度量,其中,$\hat{P}_{81,N}$表示1981年靠近焚化炉址的平均房价;$\hat{P}_{81,F}$表示1981年远离焚化炉址的房价;余类推。在微观计量经济分析中,又把这个交互作用叫作差异中的差分估计量(difference-in-differences estimator)。它代表了去掉时间差异后焚化炉址对房价的影响,所以可把它看作焚化炉址与房价的真正因与果的关系。

为了说明δ代表交互作用,把模型(1)改为
$$P = \beta_0 + \beta_1 Y_{81} + \beta_2 N + \delta Y_{81} N + u$$
也就是在(1)中加进一个时间(年份)虚拟变量Y_{81}:

$$Y_{81} = \begin{cases} 1 & \text{如果是1981年的数据} \\ 0 & \text{如果是1978年的数据} \end{cases}$$

和一个交互作用项 $Y_{81}N$。兼并使用两年的数据估计得：

$$\hat{P} = \underset{(2726.91)}{82517.23} + \underset{(4050.07)}{18790.29}Y_{81} - \underset{(-4875.32)}{18824.37}N - \underset{(7456.65)}{11863.90}Y_{81}N \quad (5)$$

$$n = 321, \quad R^2 = 0.174$$

经分析知，方程(2),(3)和(5)的系数有如下关系：

(2)和(3)的截距差：$101307.5 - 82517.23 = 18790.27$，为"时间"作用。

(2)和(3)的斜率差：$-30688.27 - (-18824.37) = -11863.9$，为"交互"作用。（或更详细地，"时间"与"远近"的"交互"。）

可见，方程(5)的估计，如果不是为了便于得到估计系数的标准误，则是不必要的。（顺便指出，$\hat{\delta}$ 的 t 统计量约为 -1.59，它相对于单侧对立假设来说，仅达到显著水平 5% 的边缘，P 值 ≈ 0.057。）

2. 正交设计与分析

下面让我们把以上的数值结果套到 $L_4(2^3)$ 正交表上来分析。（为简化计算，以千元作单位并只保留一位小数。）

表中，N, Y, YN 和 P，依次代表"远近""年份""交互作用"和"平均房价"；$m(-)$ 和 $m(+)$ 分别对应于"$-$""$+$"的列价格和，并冠以相应的"$-$"或"$+$"符号；$d = m(-) + m(+)$。

我们看到交互作用列的 $d = -11.9$，与上节的差异中的差分 -11863.9（除进位小数外）相一致。说明了正交分析和回归分析殊途同归，但这只是一个方面。

另一方面，正交分析对回归分析有所补充。第一，表之所以正交，是指任意两列的正负搭记都是均匀、完备的，即"$-,-$""$-,+$""$+,-$"和"$+,+$"都各出现同样多次（在 L_4 中，各出现一次）。

因此各列是对称的，从而交互作用列既可解释为距离远近对房价的"纯"影响，也可看作不同年份的纯影响。之所以能确认为"远近"的影响，完全是经济分析的结果。因为时间代表历史环境或趋势的变化，在一般的因果分析中都要把它去掉。如果把时间换为兴建一条地铁的前后（比方说），怎样解释交互作用便不这么简单了。这也说明为什么在微观计量经济的回归分析中往往强调因素与时间交互作用的重要性。

第二，正交表分析既有一套便于从结果推算原因的方法，同时也还有一套从原因返回到结果的计算方法。例如，在上例中，为了得到第一个组

合的结果,只须做下面的简单计算:

$$[318.1-(-49.5)-25.7+(-11.9)]\div 4=82.5$$

其中,容易看出,$318.1=\sum P$,其余的数值无非是各列 d 值冠以第一行的符号罢了。除以 4 是因为每个数值都曾是 4 项的总和,现在取其平均而已。[不难推知,第 $m(m=1,2,\cdots,4)$ 个组合的结果是将各列 d 值冠以第 m 行的符号而得到的。]但这样一来,交互作用将是 $(-11.9)/4$,而不是 -11.9。(差别来自基期或基准的选择。)当然,在不计误差的情况下,回归分析也有一套从原因推算结果的方法。(同时,正交表分析也有一套计算误差的简易方法。)然而,从方法的直观程度以至方法的不同着眼点看,对于内容的理解和掌握,回归分析和正交表分析两者是互为补充的。{例如,按照正交表,交互作用可理解为[结果$-$(平均效应$+\sum\pm$主效应)],主效应(指 N 或 Y_{81} 列的 d 值除以4)取"$+$"或"$-$",视表中相应行的符号而定。以第一个组合为例,就是

$$82.5-[318.1-(-49.5)-25.7]\div 4=(-11.9)/4。$$

当且仅当结果等于平均效应加主效应的代数和时,交互作用为零。

$L_4(2^3)$

组合	N	Y_{81}	$Y_{81}N$	P(结果)
1	$-$(远)	$-$(78)	$+$	82.5(千元)
2	$+$(近)	$-$	$-$	$82.5-18.8=63.7$
3	$-$	$+$(81)	$-$	101.3
4	$+$	$+$	$+$	$101.3-30.7=70.6$
$m(-)$	-183.8	-146.2	-165	
$m(+)$	134.3	171.9	153.1	
d(原因)	-49.5	25.7	-11.9	

3. 正交设计的价值

随着定性变量的应用由于政策(事件)分析的需要而日渐频繁,正交表,特别是二水平正交表的套用机会,越来越多了。不少经济学家(如 D. McFadden, J. Heckman 等人)认为,当今计量经济学的最主要用途在于做政策分析。政策(事件)的评价,往往涉及多个虚拟(二水平)变量的应用。多于二个虚拟(二水平)变量的情形,可套用更大的正交表如 $L_8(2^7)$,$L_{16}(2^{15})$,$\cdots$,其分析方法基本上是 L_4 的直接推广,一目了然。多于二水平

的定性变量,如多时期的时间变量,可参照多水平或混合水平正交表进行分析。

正交设计保证了误差(无论误差列或重复实验中的误差)和因素(列)是正交的。当然,这是针对真正的实验而言,即对未考虑的因素均能加以控制,使之不变。在回归分析中,最令人担忧的就是误差项与解释(自)变量之间不是正交的。其中最可能的原因是漏掉了一些有关变量。补救的办法有二:一是在回归模型中把这些可能的遗漏变量显式地加进来;二是通过抽样设计使未经考虑的那些因素的变化和已经考虑的(所观测的)因素变化成为正交或尽可能正交。严格说来,本文所引案例的两期(年)观测值并非正交。误差和因素的正交性要求数据集至少具有综列数据(panel data)的定义性质。

运用正交设计思想,改进抽样方法,减少选择性偏误,提高回归分析的质量,能使经济学更富于可实验(直接或间接)性(视观测为间接实验)。

湖北省现场统计研究会第四次代表大会概况

湖北省现场统计研究会第四次代表大会暨庆贺林少宫教授八十寿辰与2001年学术年会于2001年12月15日—16日在华中科技大学主校区召开,华中科技大学数学系承担了会议筹备与会务工作,并在财力、物力、人力上给予了大力支持。与会代表70余人。

开幕式上,第三届理事会理事长唐月英教授致开幕词并作了第三届理事会工作与财务报告。她首先代表研究会同仁向我国数理统计学家、计量经济学家、我会主要创始人、名誉理事长林少宫教授八十华诞暨从教五十年致以热烈祝贺。特别地赞誉他始终对现场统计工作与我会的发展起着推动、促进作用与影响。(节选)

多元线性回归系数的"其他情况不变"释义[①]

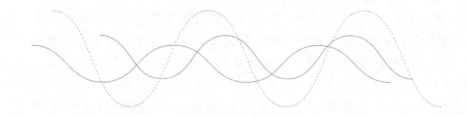

1. 引言

R. A. Fisher (1926)说过:

"大自然要求你能写出一份很有逻辑性的、思考得很完善的问卷,才会给你作出最好的回答。的确,如果你只问它单一问题,它往往在对其他问题作讨论之前拒绝回答。"

其实社会经济问题也是如此,单一问题往往得不到回答,更不用说最好的回答。

在实验设计中,这就叫做多因素、多指标问题,或者说多原因、多结果问题。

在计量经济学中,联立方程模型典型地说明经济里的问题是多原因、多结果的,它用外生变量描述原因,用内生变量代表结果。

计量经济学的发展,特别是微观计量经济学的兴起,使得单一线性回归方程中的解释变量也很强调外生与内生之分。

为说明 OLS 有良好性质的最重要的一个假定是方程中的误差项 u 和诸解释变量 $x_1, x_2, \cdots, x_k$ 不相关。为便于把 x 看作随机的,把这个形式写得强一点就是

$$E(u|x_1, x_2, \cdots, x_k)=0$$

当这个零条件期望成立时,这些 x 就是外生的;如果有某个 x_j 与 u 相关:$E(ux_j) \neq 0$,那么 x_j 就是内生的。

[①] 原文来自华中科技大学 2002 年 6 月举办的"微观计量经济学高级研讨班"上的讲稿。

倾注了大量计量学家的心血的一个问题是

$E(ux_j)=0$ 成立吗？

$E(ux_j)=0$ 不成立的理由是什么？

$E(ux_j)\neq 0$ 时怎么办？

2. 多元回归的 OLS 估计与解释

让我们考虑多元线性回归模型。

有人说，多元回归方程的 OLS 估计，解决了实验科学中保持其他情况不变(ceteris paribus)的问题。这是真的吗？

我们知道：对简单的二元回归 $Y=\beta_0+\beta_1 X+u$ 来说，β_1 的 OLS 估计量为

$$\hat{\beta}_1 = \frac{\sum xy}{\sum x^2}$$

其中

$$x = X - \overline{X}, \quad \overline{X} = \frac{\sum X}{n}$$

n 为样本大小；y 也类似地定义。$\hat{\beta}_1$ 的方差为

$$\text{var}(\hat{\beta}_1) = \frac{\sigma_u^2}{\sum x^2}$$

对多元回归 $Y=\beta_0+\beta_1 X_1+\beta_2 X_2+\cdots+\beta_k X_k+u$ 来说，也有类似的形式：

$$\hat{\beta}_j = \frac{\sum \hat{u}_j y}{\sum \hat{u}_j^2}, \quad \text{var}(\hat{\beta}_j) = \frac{\sigma_u^2}{\sum \hat{u}_j^2} = \frac{\sigma_u^2}{\sum x_j^2(1-R_j^2)}$$

其中：$x_j=\hat{a}_0+\hat{a}_1 x_1+\cdots+\hat{a}_{j-1}x_{j-1}+\hat{a}_{j+1}x_{j+1}+\cdots+\hat{a}_k x_k+\hat{u}_j$，$\hat{a}_i$ 为 x_j 对其余 $x_i(i\neq j)$ 的回归系数估计，$\hat{u}_j$ 为其残差，R_j^2 为 x_j 对其余 x_i 的相关系数的平方。可证明如下：

对多元回归方程

$$Y=\beta_0+\beta_1 x_1+\cdots+\beta_k x_k+u$$

有正规方程

$$\sum x_j(y-\hat{\beta}_0-\hat{\beta}_1 x_1-\cdots-\hat{\beta}_k x_k)=0, \quad j=1,2,\cdots,k$$

由 OLS 估计的性质，

$$\sum(\hat{x}_j+\hat{u}_j)(Y-\hat{\beta}_0-\hat{\beta}_1 x_1-\cdots-\hat{\beta}_k x_k)=0 \Rightarrow$$

$$\sum \hat{u}_j(Y-\hat{\beta}_0-\hat{\beta}_1 x_1-\cdots-\hat{\beta}_k x_k)=0 \Rightarrow$$

$$\sum \hat{u}_j(Y - \hat{\beta}_j x_j) = 0 \Rightarrow$$

$$\sum \hat{u}_j[Y - \hat{\beta}_j(\hat{x}_j + \hat{u}_j)] = \sum \hat{u}_j(Y - \hat{\beta}_j \hat{u}_j) = 0$$

由此得

$$\hat{\beta}_j = \frac{\sum \hat{u}_j y}{\sum \hat{u}_j^2}, 其中 y = Y - \overline{Y}$$

由 $\sum \hat{u}_j^2$ 和 $\sum x_j^2$ 在 OLS 估计中的关系知

$$\text{var}(\hat{\beta}_j) = \frac{\sigma^2}{\sum \hat{u}_j^2} = \frac{\sigma^2}{\sum x_j^2(1 - R_j^2)}$$

其中 R_j^2 为 x_j 对其余 $x_i(i \neq j)$ 的平方相关系数。

因此，β_j 可解释为 x_j 在去掉其余 x_i 的影响之后（成为 $\hat{u}_j$）对 Y 的影响，再把去掉其余 x_i 的影响解释为"保持其余 x_j 不变"，于是说 $\hat{\beta}_j$ 给出了 x_j 在"其他情况不变"条件下的局部效应（partial effect）；给出局部效应的每个 $\hat{\beta}_i$ 也就称为局部或偏回归系数。

一旦对多元线性回归作出 OLS 估计，就可将估计方程写为

$$\hat{Y} = \hat{\beta}_0 + \hat{\beta}_1 x_1 + \cdots + \hat{\beta}_k x_k$$

从而有

$$\Delta \hat{Y} = \hat{\beta}_1 \Delta x_1 + \cdots + \hat{\beta}_k \Delta x_k$$

若保持除 x_j 以外的 x_i 不变，便得到

$$\Delta \hat{Y} = \hat{\beta}_j \Delta x_j$$

这明显地说明了 $\hat{\beta}_j$ 的局部效应或"其他条件（变量）不变"含义。

在做简单回归时，例如需求量对价格的回归，虽然我们声称我们所考虑的是一个局部均衡，但事实上这是不真实的，因为在做抽样观测时，我们并非在保持其他变量（例如收入、其他有关商品的价格、个人嗜好）不变的情况下进行的。

3. 受控变量过少的严重性

在现代计量经济、特别是微观计量经济分析中，回归不仅是一种相关关系，还要把它当作因果关系来分析，特别是关于政策或事件评价，总想知道一项政策或计划引致了什么后果。然而，一种后果（现象）往往是多原因的，怎样把这项政策以外的原因隔离开来，多元回归的确为我们提供了重要手段。简单说来，多元回归就是在回归方程中多放进一些会影响因变量

Y 的解释变量 x_i，从而得以控制。

应引进哪些 x_i？有待于对 $E(u|x_1,\cdots,x_k)=0$ 这个关键的假定条件作细致的专业（经济）分析和统计分析。

过去常把期望值 $E(Y)$ 看成一种典型值，把误差（或干扰）u 看成对典型值的偏离，好比是一种测量误差。而现代的计量分析，主要把偏离看成未观测到的因素（unobserved factors）的作用。这些因素往往随观测单元（或对象）而异。例如，现代消费者理论注意到很多观测不到的因素在影响着消费者的选择，如果这些因素或变量与已进入方程的 x_i 相关，OLS 估计量 $\hat{\beta}_i$ 将是有偏误且不一致的。

例如，真实模型：$Y=\beta_0+\beta_1 x_1+\beta_2 x_2+u$

误设：$Y=\beta'_0+\beta'_1 x_1+v$

因此，若取 $\hat{\beta}'=\dfrac{\sum xy}{\sum x^2}$，则

$$E(\hat{\beta}'_1)=\quad \beta_1+\quad \beta_2 b_{21}$$
$$\text{直接效应}\quad \text{间接效应}$$

其中 b_{21} 为 x_2 对 x_1 的样本 OLS 回归系数。$\beta_2 b_{21}$ 代表漏掉变量偏误（omitted variable bias）。

说 $\hat{\beta}'$ 有偏误，表示平均而言简单回归估计公式是不正确的，不是偏高就是偏低。

不一致性则表示即使样本很大也纠正不了这种偏误。一位计量学家说，如果得不到参数的一致估计量，作为一种估计方法，就是没有意义的。可见当 v 中含有与被遗漏的重要解释变量（在本例中 $v=\beta_2 x_2+u$），而这个变量又与某些 x_i（在例中就是 x_1）相关时，导致了 $E(vx_i)\neq 0$ [在本例中 $E(vx_1\neq 0)$]，其后果可能是严重的。

例　　　$\log(\text{工资})=\beta_0+\beta_1(\text{教育})+\beta_2(\text{能力})+u$

但因能力难以观测，我们估计了

$$\log(\text{工资})=\beta'_0+\beta'_1(\text{教育})+v$$

由于能力通常对工资有正的影响，同时能力与教育之间往往是正相关的，因此 $\hat{\beta}'$ 有过高的偏误之嫌。

既然能力对收入有影响，在简单的回归中我们没对能力加以控制，使其保持不变，$\hat{\beta}'$ 就没有给出"保持其余条件不变"效应，也就不是一种因果关系。如果 $\Delta(\text{教育})$ 作为一种政策，则我们并没有达到政策评价的目的。

$$\Delta\log(\widehat{\text{工资}})=\hat{\beta}'\Delta(\text{教育})$$

4. "代理"与"工具"

考虑方程：$\log Y = \beta_0 + \beta_1 X + u$，$X = $教育，$Y = $工资。

怎样解决 $E(ux) \neq 0$ 的问题？

方法之一是从误差项 u 中把一些有份量的变量（比如说 w）调遣出来，如果不难观测就用 OLS 多元回归加以控制；如果难于观测，则可设法找到代理变量（proxy），比如说 z'，要求 z' 不仅可以观测，且与难于观测的变量 w（如能力）有尽可能高的相关关系，同时与已包含在方程中的解释变量（如教育）有尽可能低的相关关系，然后进行 OLS 估计。

另一方法是用工具变量 IV 比方说 z 代替 x，但要求 z 与 x 有尽可能高的相关而 z 与 v 则有低到零的相关。

由于 w 被包括在 u 中，看来，对工具变量的要求和对代理变量的要求适得其反。

须知，工具变量 z 是作为 x 的替身，而代理变量 z' 是作为 w 的替身的，对两者的要求相反，并不奇怪。

拿工资、教育与能力之间的回归来说，因能力被包含于误差 u 之中，对工具变量 z 的要求是 $\text{cov}(z,x)$ 愈大愈好；$\text{cov}(z,v)$ 则要小到零。事实上要求 $\text{cov}(z,v) = 0$。另一方面，对代理变量 z' 的要求则是 $\text{cov}(z',x)$ 愈小愈好，而由于能力包含在 u 中，$\text{cov}(z',v)$ 愈大愈好［事实上必须 $\text{cov}(z',v) \neq 0$］。

确实，对工具变量 z 的要求

$$\left. \begin{array}{l} \text{cov}(z,v) = 0 \\ \text{cov}(z,x) \neq 0 \end{array} \right\} @$$

足以识别 β_1；意谓可在条件 @ 下一致地估计 β_1；可以用总体矩表达参数 β_1，并用相应的样本矩估计之。

现阐述如下：

$$Y = \beta_0 + \beta_1 X + v$$

取 z 与 y 的协方差得

$$\text{cov}(Z,Y) = \beta_1 \text{cov}(Z,X) + \text{cov}(Z,v)$$

因 $\text{cov}(Z,v) = 0$ 故

$$\beta_1 = \frac{\text{cov}(Z,Y)}{\text{cov}(Z,X)}, \quad \text{cov}(Z,X) \neq 0$$

其 IV 估计量将是

$$\tilde{\beta}_1 = \frac{\sum zy}{\sum zx}$$

大数定律告诉我们 $\tilde{\beta}_1$ 是一致估计，但在小样本中，$\tilde{\beta}_1$ 可能有较大偏误。

β_0 的 IV 估计和 OLS 估计类似，也就是 $\tilde{\beta}_0 = \overline{Y} - \tilde{\beta}_1 \overline{X}$

在同方差假定之下，$\tilde{\beta}_1$ 渐近地为正态分布：$\tilde{\beta}_1 \sim N\left(\beta_1, \frac{\sigma^2}{n\sigma_x^2 \rho_{xz}^2}\right)$

例　对已婚妇女的教育回报估计

模型：$\log(工资) = \beta_0 + \beta_1 (教育) + u$

OLS 估计：$\log(工资) = -0.185 + 0.109(教育)$

$\qquad\qquad\qquad$ (0.185)　(0.014)

$$u = 428, \ R^2 = 0.118$$

$\tilde{\beta} = 0.109$ 意味着每多受一年教育约增加 11% 的回报。

以父亲教育作为教育的工具变量时，可先做相关检验：

$\qquad\qquad$ 教育 $= 10.24 + 0.269$（父亲教育）

$\qquad\qquad\qquad$ (0.28)　　(0.29)

$$n = 428, \quad R^2 = 0.173$$

也就是说父亲教育解释了样本中教育变导的约 17%，然而这是显著的。

IV 估计：$\log(\widehat{工资}) = 0.441 + 0.59(教育)$

$\qquad\qquad\qquad$ (0.446)　(0.035)

$$n = 428, \quad R^2 = 0.093$$

IV 估计的教育回报为 5.9%，约合 OLS 估计之半，似乎说明了 OLS 估计确有偏高之嫌。（须知，这仅是一个样本结果。）另外，IV 估计的标准误比 OLS 估计的大得多，差不多大到两倍半，这也是符合理论的。

5. 讨论

5.1　边际效应

对偏回归系数总能做出"其余情况不变"的解释吗？

二次项：

$$Y = \beta_0 + \beta_1 X + \beta_2 X^2 + u$$

例如 $Y = 工资$，$X = 工作经验$。说改变 X 而保持 X^2 不变，将毫无意义。对 $\hat{Y}=\hat{\beta}_0+\hat{\beta}_1 X+\hat{\beta}_2 X^2$，近似地有 $\Delta\hat{Y}=(\hat{\beta}_1+2\hat{\beta}_2 X)\Delta X$ 或 $\Delta\hat{Y}/\Delta X=\hat{\beta}_1+\hat{\beta}_2 X$。

关于边际效应的假设可以是
$$H_0:\beta_1+2\beta_2 X_0=0,$$
$$H_A:\beta_1+2\beta_2 X_0<0, \quad X_0 \text{ 为某给定值}。$$

这相当于问 X_0 是否是转折点。为此，还需要给出对 $\beta_1+2\beta_2 X_0$ 的类似于对 β_j 的"其他情况不变"估计和解释。

交互作用：

例如，房价 $=\beta_0+\beta_1$ 平方米 $+\beta_2$ 卧室间数 $+\beta_3$ 平方米·卧间 $+\cdots+u$

卧室间数（简称卧间）对房价的局部效应为：
$$\frac{\Delta \text{房价}}{\Delta \text{卧间}}=\beta_2+\beta_3 \text{平方米}$$

检验 $H_0:\beta_3=0$ 显然是有意义的（这相当于检验某种斜率是否不因平方米而变），但能说平方米和卧间都保持不变而其乘积在变吗？

当然，还可以考虑更高次的回归函数，比如三次式的总成本函数。

5.2 受控变量越多越保险？

似乎受控变量越多，实现"其他情况不变"的可能性越大。因而支持了所谓"从一般到特殊（from general to specific）"的建模方法。

问题不在于多或"从多到少"，而在于过多或过少。和受控变量过少相比，过多的控制变量仍保持有 OLS 估计量的无偏性和一致性。所以说，以统计学的角度看，过多没有产生像过少控制变量那么严重的后果。但从经济学的角度看，就失去了理论上的简单概括性；不能以最少概括最多。

另外，究竟什么是需要保持不变的"其他情况"？H. J. Bierens 等（2000）在计量经济学原理论坛上的一篇论文中说："在经济理论中和在计量经济学中假定'其他情况不变'的主要差别在于，在经济理论中，即使世界现况（state of the world）尚未给定，事实上则总是附加上这个条件的。而另一方面，在计量经济学中，这个条件却被用来设想出一个实验，以利于解释估计的结果。"

经济学家并不完全知道"其他情况"指的是什么，但计量学家却利用它来对估计结果做"其他情况不变"的解释。由此导致了 Bierens 等对"其他情况不变"的真实性进行概率计算。

一些受控变量明显过多的例子还是不难列举的，例如，一项为控制酗酒驾车的现象而加重酒精饮料的课税政策,目的在于减少饮酒量,如果在回归方程中既放进一个课税变量,又考虑一个饮酒量,那么逻辑上就是矛盾的。又如在回归方程中既放进一个限速变量,而同时又考虑一个平均驾驶速度变量也是矛盾的。

5.3 t 值小而 R^2 值大的估计方程没有用吗?

t 值小的回归系数似乎很难对它做"其他情况不变"的解释。

t 值小而 R^2 值大的估计方程,常常是一种多重共线性现象。受控的变量考虑的多,虽然防止了漏掉重要的"其他情况",却添加了多重共线性的危险。当然,不能因为担心多重共线性而放弃可能重要的解释变量。

明显共线的变量是应该避免的。例如,既然用了父亲的教育作为教育的"工具",就不必考虑母亲的教育。这并不是说母亲的教育不可以作为"工具",而是因为父亲的与母亲的教育有较高度的相关。

一个回归系数的 t 值小,对其做"其他情况不变"的解释,确实是个问题。但不能就此说相应的变量不重要,因为在 t 小而 R^2 大的情况下,我们应该检验的常常不是 $H_0:\beta_j=0$,而是 $H_0:$ 诸 β_j 的某个(线性)组合$=0$。预测时也要注意到只有对诸 x_j 的某个组合而言才有意义。这也构成对某一 $\hat{\beta}_j$ 的"其他情况不变"解释的一种挑战。

5.4 只关注某些 β_j 或它们的某种组合

往往,研究者只对某些 β_j 或其线性组合感兴趣。例如,

(1)政策研究者关心财产税的变化对房价的影响,例如,OLS 估计得

$\widehat{房价}=0.171+7.275$ 减税 $+0.547$ 平方尺 $+0.00073$ 中等收入

t 值:　　　(2.97)　　　(2.32)　　　(1.34)

$\qquad +0.0638$ 房龄 -0.0093 距离中心城市 $+0.857$ 评估价

　　　　(3.26)　　(2.24)　　　　　　(1.80)

被调查城镇数　$n=64, R^2=0.897$

假定减税额度一直维持不变,每减税 1 元,房主今后节省缴税额的现值为

$$\frac{1}{1+r}+\frac{1}{(1+r)^2}+\cdots=\frac{1}{r}$$

鉴于抽样调查期间的利率 r 在 12%~15% 之间,故认为所估计的减税的"其他情况不变"效应是合理的;无论从统计意义上或经济意义上看,估计值 7.275 都相当不错。

(2) 房产评估机构关心评估价格是否与实际价格一致。

模型：$\log(房价) = \beta_0 + \beta_1 \log(评价) + \beta_2 \log(平面积) + \beta_3 \log(平方尺) + \beta_4 卧间数$

OLSE：$\log(房价) = -0.34 + 1.043\log(评价) + 0.0074\log(平面积)$

s.e　　　　　　　(0.972)　(1.51)　　　　　(0.0386)

　　　　　　　$-0.1032\log(平方尺) + 0.0338 卧间数$

　　　　　　　　　(0.1384)　　　　　　(0.0221)

$$u = 88, \quad R^2 = 0.773$$

所感兴趣的待检验的虚拟假设将是 $H_0: \beta_1 = 1, \beta_2 = \beta_3 = \beta_4 = 0$。为此，我们还必须估计

$$实价 - 评价 = \beta_0 + u$$

以便进行方差分析，查看在给定 H_0 下回归的边际效应。

(3) 离散选择。

如，交通模式的选择概率 $= f($供方因素：路线、定价、服务、速度、安全、舒适；需方因素：经济、社会、心理、个人历史方面$)$

可能最受关注的解释变量：价格、路线、消费者的社会、经济地位等。

6. 因果效应与综列数据

在工资、教育与能力一例中，作为能力的代理，人们曾考虑 IQ（智商），是否还可考虑 AQ（管商）、FQ（财商）等因素呢，也许这些商数和能力的测量一样的模糊，或者我们知道有影响着工资的潜在因素而不知道把它叫作什么。为了选择一个"代理"这个模糊的潜在因素，选用滞后应变量未尝不是一个好主意。例如，过去的工资也许代表一种潜在的能力。更好的说明例子也许见于以下模型：

$$犯罪率 = \beta_0 + \beta_1 失业率 + \beta_2 执法开支 + \beta_3 犯罪率_{-1} + u,$$

这里把前一期的犯罪率$_{-1}$加进方程中，以"代理"原来误差项中含有的影响犯罪率的所有潜在因素。这样做就相当于把影响当前犯罪的历史因素放进方程中加以控制，从而可望算出执法开支这个"政策变量"的更为准确的因果效应，参见表1。

表1 应变量 $\log($犯罪率$)$

自变量	1	2
失业率	$-0.29(0.032)$	$0.009(0.20)$

续表

自变量	1	2
log(执法开支)	0.203(0.173)	−0.140(0.109)
log(犯罪率$_{-1}$)	—	1.194(0.132)
截距	3.34(1.25)	0.067(0.821)
被调查城市数	46	46
R^2	0.057	0.680

容易猜想,如果还知道过去的失业率和执法开支,将能更好地分析导致犯罪的因果关系,这有赖于收集更多的数据,特别是所谓"综列数据(panel data)"。

7. 结束语

在实验科学中要求除一个变量外其余变量均保持不变的主张早已被人们否定。取而代之的析因实验(factorial experiment),要求同时变动多个因素。在多元回归中,我们也是同时考虑多个因素的变动,并不是孤立地只变动一个因素。其实,这是经济社会的自然现象,本讲所讲的"其他情况(或因素)不变"并非一种实验(观测)方法,而是为了寻求对实验(预测)结果的解释。事物总是多原因、多结果和互为因果的,例如在上例中,如果执法开支增加,使执法力度更大,犯罪行为便会受到更多的抑制。反过来,犯罪率一旦减少,执法开支也会随之减少。这样就把犯罪与执法开支的因果关系颠倒过来。所以说,需要用一个多方程联立的模型加以描述、估计和解释。结果是,通过多个外生变量的变动,同时估计多个内生变量的变化。只有这样,才能得到所研究问题的较好答案。

微观计量经济学中的实验设计法
——期望"计量经济(学)设计与分析"的出现

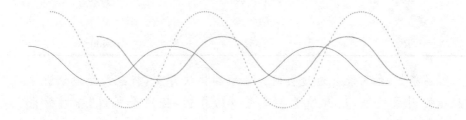

内容提要：本文论述了实验设计思想在微观计量经济分析中的重要作用。从实验设计三大原理——局部控制、随机化和重复——出发，重新诠释微观计量经济学（特别是"因果链"分析）中有关方法论的问题，预期实验经济学和微观计量经济学将通过实验设计的思想方法而互相促进，并期望"计量经济（学）设计与分析"一类读物的出现。

关键词：实验设计；因果链；回归模型

Methodology of Experimental Design in Micro-Econometrics: Expecting the Appearance of Micro-Econometric Design and Analysis

Abstract: This paper argues for the importance of experimental design in micro-econometric analysis, re-explains the methodology of microeconomics (especially for causal analyses) and expects the appearance of many writings on micro-econometric design and analysis.

Key words: experimental design; causal chain; regression model

不要认为计量经济中的数据是非实验性质的就没有实验设计可言。其实，实验设计不仅包括数据的获得（如来自实验室或来自社会调查），还涉及对数据的要求，估计的方法等等，而估计方法自然和数据的性质有密

---

① 原文发表于《统计研究》2007年第24卷第2期。

切关系。在微观计量经济学中,为使模型估计能个别地适合微观经济单元(如个人、厂商或地区),谈如何设计数据就没有什么奇怪的了。

实验设计曾是数理统计学的一个分支,它的创始人 R. A. Fisher 早已定下实验设计的三大原理,即局部控制(local control)、随机化(randomization)和重复(replication),至今被科学实验工作者奉为圭臬。这三大原理的核心在于控制,对计量经济学来说尤其如此。为了做好政策评价或事件研究,需要找出变量之间的因果关系。这时往往要求设法控制除一个因素(变量)外其余因素均保持不变。最小二乘多元回归之所以被普遍应用,其主要原因之一也就在此,即用统计方法保持其余因素不变。(至于"其余因素"具体指什么因素,既是常识问题,但也可能涉及较深层的理论,如由通胀预期扩大的菲利普斯曲线)。本文的论述首先把"其余因素"看作常识问题。

在开展讨论实验设计三大原理,特别是局部控制或简称控制之前,不妨比较一下 D. McFadden(2000 年诺贝尔奖获得者),在《怎样量化环境损坏或改进的经济价值》一文中所提出的评价实验三原理:(1)选择可与"处理(treatment)组"相比较的"对照(control)组"。(2)如果混杂变量(confounding variables)不能保持不变,则需设法(用统计方法)校正或通过随机化而使其中性化。(3)利用已知的解答,作为基准。

以上第(1)条是要设计不受处理的对照(或控制)单元;第(2)条是要控制住起混杂作用的变量(或因素)使其保持不变,或通过随机化以抵消其影响;第(3)条相当于重复别人的或自己过去的成功经验,看成功的结果是否重现以及误差如何。通过重复而减少误差可见 D. McFadden 的评价实验三原理与 R. A. Fisher 的实验设计三原理一脉相承。"控制"仍然是实验设计的核心。

1. 控制什么?怎样控制?

对同一解释变量在简单回归和多元回归中用最小二乘法估计的回归系数,一般说来结果是不相同的,在含不同解释变量的多元回归中,这个估计值也是不同的,除非数据矩阵是正交的(例如按正交表安排试验)。在一个线性多元回归中应含多少个解释变量就是一个控制问题,把某一个解释变量放到回归方程中,根据最小二乘法原理,就说是对该变量作了控制。这正是把解释变量叫做控制变量的原因。例如考虑教育的回报,如果把工龄作为另一解释变量写进回归方程,就说明对工龄的变异作了控制;最小二乘法的使用就是用统计方法保持了它的不变。同样,再放进其他变量,其含义也是类似的。为了说明回报的变化是由于教育的变化这种因果关系,显然有必要对同时影响回报的变量(如工作经验)加以控制,使其(在统计上)保持不变。反之,如果不把工作经验放进回归方程,则说回报(结果变量)被工作经验(待控变量)混杂了。

  当解释变量作为一个政策变量时,人们都想知道这个政策的效果,这个效果将体现在受政策影响的人和不受政策影响的人在政策实施后的差别,那么这两部分人需要有可比性,也就是这两部分人除了受和不受政策影响外,在其他多个方面都应该是相同的(或者说越相似越好),例如收集孪生兄弟的有关数据来控制。因此,控制好调查(或观测)对象的分配,使得"处理组(受政策影响者)"和"对照组(不受影响者)"有可比性,是重要的。这个问题看来简单,具体做起来并不容易。有许多复杂性和灵活性需要具体分析。

  处理组与对照组的设置来自生物、医药试验。一种新药或疗法效果,要在处理组与对照组之间反复进行比较才能得出结论。同样,一种教育政策或培训方法效果如何?一条法律的颁布,一项措施的执行,一桩事件的发生在哪些方面发生影响?都需要在处理组和对照组之间进行比较。用回归的术语来描述,可用一个虚拟变量的取值来区分处理组和对照组。例如,对受事件影响者,虚拟变量取值1,代表处理组;而不受事件影响者取值0,代表对照组。金融市场常受某一事件的影响而出现新的波动,那么在事件发生前(后)的市场情况可用0(1)表示,代表对照组(处理组)。考虑到一个化工厂的建立会影响附近房价,那么靠近该厂的房屋为1,远离的为0。但这些房子除远近不同外,其他方面应尽可能相同,或者通过目的抽样使之相同,就是控制(组)的涵义。在农业实验中为了研究新肥料或新耕作方法,常设对照组以资比较,要求在同一块土地上(代表同样的土质,同样的其他自然条件),按原肥料或原耕作方法种植,作为对照,已是司空见惯的事。但把对照组的概念移植到计量经济学中来,还是在微观计量经济出现后(特别是 2000 年诺贝尔经济学奖授予 J·赫克曼和 D·麦克法登之后),使实验设计有了用武之地,才逐渐引人注意。

2. 不可观测变量的控制

  为了研究事物的因果关系而需要保持不变的因素不一定是可观测的或可观测而极难于准确观测的。例如影响个人收入的因素,不仅有教育程度,还有工作能力、传统势力,等等。而能力、传统都是不易观测的,如何通过统计方法保持能力、传统等不变,以便估计教育的回报,将是一个挑战性问题。又例如,一个人的睡眠时间或工作时间和他的难于观测的生理和心理状态有关,利用面板数据(panel data)的固定效应法可以提供一个很好的控制方法。

另一个方法是寻找替代变量(称工具变量),设法把不可直接观测的因素和可观测的因素联系起来,用可观测的因素替代不可观测的因素。例如用智商替代能力(这自然要求智商与能力相关越大越好,另一方面智商还需与原回归模型中的误差项无关)。替代变量的寻找也可看作控制的一种近似方法,两者相关程度越高,控制的程度也越高。一个较常用的方法是用滞后应变量作为代理(proxy)变量(有别于工具,但这里不去考虑这种技术上的细节)当然也有其局限性。

当待控变量难于直接观测时,利用待控变量与关注(解释)变量的正交性选取观测对象,也未尝不是一种可参考的方法。例如为了找出教育(关注变量)的回报,由于教育与家庭背景有关,并且家庭背景(待控变量)是一个难测的变量,于是可通过设计使每种教育水平的人都与各种家庭背景组合同样多次,以观测这种组合的回报。这种回报从原理上讲将不受家庭背景不同的影响。收集孪生兄弟的有关数据做配对实验,也是消除出身背景差异的一种方法。另外,还可考虑在回归方程中引进一个滞后内生变量作为控制变量,等等。

3. 非严格的可估计函数[①]

L. R. Klein 和 A. S. Goldberger (1962) 曾在 An Economic Model of the United States, 1929—1952 一书中试图对美国经济拟合如下回归模型:
$$Y_i = \beta_1 + \beta_2 X_{2i} + \beta_3 X_{3i} + \beta_4 X_{4i} + u_i$$
其中,Y 代表消费,X_2 代表工资收入,X_3 代表非工资、非农场收入,而 X_4 为农场收入。因预料时间序列 X_2, X_3 和 X_4 高度共线,故通过横截面分析把 β_3 和 β_4 估计为 $\beta_3 = 0.75\beta_2$ 和 $\beta_4 = 0.625\beta_2$,利用这些估计,他们重新建立消费函数如下:
$$Y_i = \beta_1 + \beta_2(X_{2i} + 0.75X_{3i} + 0.625X_{4i}) + u_i = \beta_1 + \beta_2 z_i + u_i$$
其中,$z_i = X_{2i} + 0.75X_{3i} + 0.625X_{4i}$;可用这个新建的模型去拟合对 Y,

---

[①] 严格意义下的可估计函数是指完全共线性意义下的可估计函数。

X_2, X_3 和 X_4 的观测数据,并估计 β_1 至 β_4。

可从另一角度把估计式 $\beta_3 = 0.75\beta_2$ 和 $\beta_4 = 0.625\beta_2$ 解释为
$X_3 = $ 常数 $K_3 + D_3 X_2 = K_3 + 0.75 X_2$ 和 $X_4 = $ 常数 $K_4 + D_4 X_2 = K_4 + 0.625 X_2$
其中 $D_3 = 0.75$ 和 $D_4 = 0.625$ 由横截面估计出。

重要的是,要看到为了预测,不能个别地(自由)变动 X_2, X_3 和 X_4 以观测其对 Y 的影响,而是要根据组合 $X_2 + 0.75, X_3 + 0.625, X_4$ 变动或微调 Z 以预测 Y 的变化,这样才有比较准确的预测。之所以有 Z 这个组合,是因为 β_2、β_3 和 β_4 有如上的非严格可估性约束。怎样找到这个约束就是个控制的问题。比如,氮、磷、钾肥都是农作物生长所需要的,因此很难确定其中一种肥料的效应,它们总是按一定比例联合地起作用的,正好比本例中的 X_2, X_3 和 X_4 那样。

用主元法(principal components)寻找最显著的自变量组合也许是个好方法,并不像一些人所认为的,这样的组合缺乏经济意义,难于作经济解释。正好相反,这样做有着相当明显的经济意义,即为了达到最好的经济效益,往往需要多个自变量或控制变量一定的组合,问题仅在于主元法未必能准确地找出最优的组合。由于不可避免的抽样和观测误差,要求能灵活地看待并处理由主元法找出的各个解释能力由大到小排列的正交组合成分。至于主元法受测量单位的影响这一困难是可以克服的。重要的是,作为生产函数的投入,诸生产要素确实要求有一定的"最优"组合;为了培植农作物的生长,各种施肥量要有一定的比例,还要配合以一定的耕作方法。作为政策变量,各相关政策也要有一定的配合。多重共线性数据的利用,关键在于寻找控制变量的"最优"组合,从而得以估计非严格意义下的"可估函数"。

4. 实验经济学

经济学中仍然有些问题是可以用实验室数据去有效地解答的。凡是用到实验室数据去解释人们的经济行为,从而形成或证实经济理论,建议或评价政策的学问,都属于实验经济学(experimental economics)的范畴。由于经济和金融领域中的实验设计不断改进,使实验室条件逐渐逼近于真实,实验的结果能为经济学者提供富有价值的洞见,并在现实中得到了应用。实验经济学也因此迅速地受到重视,以至 2002 年诺贝尔经济学奖授予了对实验经济学最有贡献的人 Vernon Smith。

实验经济学能做出一些什么独特的贡献呢?

例如,为了观察和分析股民对待风险的态度而进行实验室实验,必须在实验室中建立一种风险的环境,迫使参与试验的人在一些环境中作出决策并付诸行动。通过改变风险的性质和相应的回报,看他们怎样作出反应,就有可能了解股民是怎样在现实的股市中采取行动的。这里最重要的是:能够在实验室中模拟真实的风险环境。一些竞价买卖活动如拍卖,是比较容易在受控制的实验室环境中进行模拟的,因此这类活动的实验结果可以为政府怎样设计拍卖提供参考。但更多的社会经济问题如教育政策的制定、医疗保险制度的建立,或对农产品的贴补,等等,其真实环境就很难在实验室中模拟了。

一般说来,带有博弈性质的行为,尤其那些能较简单地作出决策的情况,是比较易于实验室模拟的。最典型而简单的问题莫过于"囚徒的困境",它的描述虽然简单,却在寡头竞争经济中得到了有价值的应用。

实验经济学通过实验室的实验数据,不但可以看到社会经济行为达到均衡状态的过程,更重要的是对经济人的非理性行为提供了一种合理的解释。在实验室里能更好地做到"除一个因素外其余因素保持不变",所以更有利于找出变量之间的因果关系。当然,这仅是"有利于",要找出真正的因果关系仍非易事。可以说,实验经济学为计量经济学出了一臂之力,仍期待着它们两者在探索事物因果的道路上有进一步的配合。

5. 期待"计量经济(学) 设计与分析"的出现

A. S. Goldberger 和 J. N. Wooidridge 等人说得很干脆:通过经济理论分析而得到的所谓结构模型或结构关系式就是考虑因果关系的。可见因果关系在计量经济学中的重要性,为了找出因果关系,需要实验设计,特别是控制方法。

综上所述,计量经济设计的主要问题包括有:(1)除一个因素外,其余因素如何最好地保持其不变。(2)对有高度交互作用的多个因素,如何最好地确定其组合中的比例关系。(3)对已成为历史的时间序列数据(应该说综列数据)如何最好地现成地加以利用;为了寻找因果关系,首先要明确是否有有关变量之间的稳定的长期关系。

第五部分
国际交流

为紧跟学术前沿,林少宫教授注重国际交流。改革开放不久,林老师访问了麻省理工学院(2次)、明尼苏达大学和伊利诺大学等美国名校,通过演讲和交流,与对方同行建立了良好的学术关系并获赠学科前沿读物,如麻省理工学院费歇尔(F. M. Fisher)教授1981年当面赠送的《Identification Problems in Econometrics》专著。1987年夏,林教授和张培刚教授等与美国学者共同在武汉华中工学院举办了"中美经济合作学术会议",会后合编了会议论文集《China's Modernization and Open Economic Policy》(包括林教授的论文"Aspects of Technology Transfer:China's Experiences"),于1990年由美国JAI出版社出版。2000年夏,林教授联系海外校友,在华中科技大学成立了数量经济与金融研究中心(着重于经济学前沿发展研究及现代化、全球化经济与金融学人才培养),并以该中心为阵地,聘请国际知名学者,包括诺奖得主赫维茨(L. Hurwicz)、麦克法登(D. McFadden)和赫克曼(J. Heckman)等,计量经济学大师拉丰(J. Laffont)和周至庄等,来这里交流、讲学,使我们的师生对其学科在国际上的最新进展有所了解,其视野也得到开阔。林教授坚持跟踪现代经济学前沿的原因是,他认为在经济学研究方面,信息与基本功同样重要,没有足够的信息,就不可能有学术上的创新。

1986年11月,林教授参加在日本福冈九州大学举办的第二届中日统计讨论会,发表论文"有限总体推论应该遵循不同的原则吗?"。

1981年和1985年,应美国罗格斯—新泽西州立大学亚洲经济研究中心的邀请,林教授和张培刚教授参加了在美国召开的第一、二届美国与亚洲经济关系会议,先后发表两篇论文:"中国的经济调整与外贸前景""中国的现代化:稳定、效率和价格机制"。

上述的论文中,有3篇被本章收录。

Aspects of Technology Transfer: China's Experiences[①]

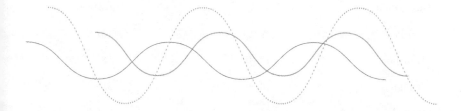

1. Economic Growth: The End Product of Technology

In recent years, especially in the 1970s, some attempts have been made to measure technological progress and its relation to economic growth in China. While opinions differ as to how much technological progress contributes to economic growth, it is a consensus that technology is the driving force behind growth in a modern economy.

In the developed countries, expenditure on research and development (R&D), number of patents, and innovations in products or processes of production have variously been used as indices of technological changes. In the less developed or developing countries, foreign investment may serve as a convenient proxy for advancement in technology. Most of these countries depend on imported technology, and these imports come with various forms of foreign investment inflows.

The present investigation begins with a discussion of the relationship between direct foreign investment (DFI) and the profit data for the 14 coastal open cities in China, in 1986, granted that profit makes a positive, and under certain circumstances, important contribution to economic growth.

---

① 原文发表于 Research in Asian Economic Studies, 1989(2): 239-251.

Table 1. Fourteen Coastal Open Cities of China: Profit and DFI in 1985

City	Y	X
Beihai	23.62	180
Zhanjiang	25.23	2,248
Guangzhou	38.77	10,216
Fuzhou	29.14	1,512
Wenzhou	17.45	3
Ningbo	40.66	359
Shanghai	62.33	10,242
Nantong	28.71	131
Lianyungang	15.31	50
Qingdao	38.62	230
Yantai	20.62	170
Tianjing	33.77	3,251
Qinhuangdao	19.07	141
Dalian	31.66	1,432

Notes: Y = Profit and sales tax per hundred yuan initial value of fixed assests (yuan).

X = Direct foreign investment (millions of yuan)*

$$Y = 24.942 + 2.512 \cdot 10^{-3} X$$
$$(6.873 \cdot 10^{-4}) \quad r = 0.726$$

* One might consider using lagged values of X in estimating the regression model. However, nonavailability of data precludes such experiments.

Source: NSBC (1986).

In a linear regression model, variations in profit plus sales tax per hundred yuan intitial value of fixed asset is explained by DFI (millions of yuan). The regression coefficient estimated is 2.512×10^{-3}, significant at 1% confidence level. The correlation coefficient is 0.726 (see Table 1) (for data sources see NSBC, 1986: 75-76). Interpretation of the above regression results relative to the role of technology, however, involves a number of difficulties. First, the data of the variable, profit plus sales tax, referred to as profit for subsequent discussions, are averages of all kinds of investments for each city, and any ad hoc adjustment of the available series to relate it specifically to DFI would be arbitrary. Second, no data on DFI as percentages of total investment for each city is reported. In any case, profit would not necessarily be proportional to the amount of undistributed corporate profit that would be retained and reinvested.

Further, availability of skilled labor, matching magnitude and standard of the technological component for the labor input, state of relative industrial advancement of a given city in the sample, and last but not least, forms of ownership of the enterprise may be responsible for differences in profits. Be that as it may, with all its limitations, the simple regression model (Table 1) points to the positive role of DFI-induced technology transfer in China's economic growth.

Statistical analysis, based on data from diverse ownership forms of direct foreign investments (DFI) in 324 cities all across the country with varying population sizes may lead some additional support to the above finding. Data indicate that profits (provided but not realized until the end of the year) per hundred yuan initial value of fixed asset vary between ownership forms—with state-owned enterprises doing the poorest, collectively-owned enterprises doing a good deal better, and other ownership forms including partnership and self-employed doing the best, in terms of earning profits. When cities are classified by popualtion size, the profit ratio is positively correlated with the population size of the city (see Table 2) (NSBC, 1986: 73-74). It is generally true that highly populated cities have greater abilities, both in terms of industrial facilities available and historical reputation as centers for industrial activities, to attract foreign investors.

Table 2. Profit and Sales Taxes
per Hundred Yuan Initial Value of Fixed Assets
(324 cities in China by Forms of Ownership and by Population Size, 1985)

	A	B	C	D	E	F	G
State-owned Enterprises	22.4	25.3	39.1	27.4	22.2	18.2	17.9

续表

	A	B	C	D	E	F	G
Collectively-Owned Enterprises	35.0	42.0	60.9	46.6	40.0	34.3	26.3
Other Ownership	45.8	47.9	60.0	43.0	50.6	39.6	23.9

Notes: A = For the country as a whole,
  B = Averaged over 324 cities
  C = Cities with population above 2 million
  D = Cities with population from 1.5 to 2 million
  E = Cities with population from 500 thousand to 1 million
  F = Cities with population from 200 to 500 thousand
  G = Cities with population under 200 thousand.
Source: NSBC (1986).

  When we move from cross-section to time series data, impetus provided by imported technologies to economic growth is more evident. A typical example is provided by rapid development of Dongguan County, situated between Guangzhou and the Hong Kong-Shenzhen economic corridor. Taking advantage of its geographical location, Dongguan was able to nearly redouble its gross industrial and agricultural output over the past seven years through technology transfer in the form of compensation trade and processing of foreign materials according to sample specifications or trademarks provided(Guangming Daily, May 28, 1987).

  Different forms of ownership mean different responsibilities assumed by the entrepreneurs. Partnerships and self-employed establishments assume the highest order of responsibility for their loss or profit. The burden of responsibility seems to be in a descending order as we move to collectively-owned enterprises and state-owned enterprises, in that order. The higher the responsibility(risk), the greater will be the reward and the profit ratio. There appears to be a positive correlation between risk and profit earning, a plausible economic hypothesis. A case in point is the fast growing economy of Jiangsu Province which includes Shanghai City. During the seven years from 1976 through 1983, Jiangsu was able to double its gross industrial and agricultural output, resulting in a per capita income of 1,200 yuan in 1983. This achievement is attributed to (Deng,1987):

(1) Internal transfer of skilled labor provided from the pool of retired workers from Shanghai City;

(2) Adoption of collective ownership by a large number of medium and small size enterprises.

Technology and collective ownership are considered the two pillars of sustained economic growth.

2. Technology Transfer: Wage, Employment and income Distribution Profile

It is not true to say that there is always automatic full employment in a socialist economy, though everyone in China would be offered a job of some description. New technology is beneficial to employment both for creating job opportunities to the unemployed and raising the productivity, and thus earnings of those who are employed. It is important to note that migration of labor as a factor of production in China is limited. For one thing, availability of residence in cities is restricted. Even in the same locality, workers find it difficult to change jobs from one trade to another. It follows that job opportunities and wage earnings of workers are very much localized.

Since technology is usually acquired or allocated, directly or indirectly, through state's planning, and not quite by open competition, enterprises with inferior technology will not be forced out of business, and workers in an enterprise with advanced technology are not rewarded with higher wages. With a view to minimizing the effect on income distribution, state regulations provide that remunerations for those employed in a joint venture with foreign participation or in a wholly foreign-owned enterprise, equipped with advanced technology, should not be more than 120% of the prevailing wage rates in comparable job descriptions. In the process, technology-labor substitution effect, if any, will be concealed in a socialist economy, at least for the short period. In China, lay-off of workers following introduction of new technologies is indeed a rare occurence.

As a corollary, management of new technologies in China often find the problem of laying-off inefficient workers as difficult as that of recruiting new efficient ones. Management-labor disputes are ususally settled by a process of persuasion and/or education. Technology-induced competition is nowhere keen. Wage rate and employment terms are inflexible. Further, benefits form improved technology are expected to go beyond the individual enterprises who acquire/or "own" it, and permeate through the economy in general. All the people of China will be its beneficiaries.

However, over the past several years a marked redistribution of income has occurred. This is regarded as a disturbing factor insofar as overall social stability is concerned. Initiation of entrepreneurial autonomy in the responsibility system of production management, which may be viewed as a software of technology (managerial technology), both in the countryside and in urban enterprises, has been responsible for making some people richer ahead of others. The resulting redistribution of income and wealth is considered a positive economic result, only if it is in conformity with the state's macroeconomic policy. The responsibility system includes an element of a free enterprise market economy insofar as it offers to relate motivations for individual gains with increased production and productivity gains. It has helped some farmers and self-employed businessmen become relatively richer. For its well-defined macroeconomic goals, the state has introduced periodic income adjustments.

An adverse effect that accompanies importation of technology relates to the price level as it tends to rise. High quality products (new designs and varieties) and superior-grade services, as they come off of the production lines of new technologies, find a market demand, despite their relatively higher prices. For a period of time, especially for 1985-1987, a lifestyle of high-income-and-high-consumption, though not a part of China's adopted macroeconomic goals, came to prevail, and the general price level rose to an abnormal high.

Establishment of "special" economic zones and coastal "open" cities may be viewed as steps to reconcile economic growth with social stability. As economic measures taken for these zones and cities are not applicable elsewhere, the effect of technology transfer on wage, employment, and income distribution, and thus on the social system and ideology, is effectively contained in specific locatlities. Changes, as may very well be desirable in the long run, are allowed to take place only in moderation. Gradual absorption of new technology into China's production system is the norm.

3. Short-Run Versus Long-Run: The Familiar Economic Debate

For a private enterprise system, profit as a measure of success merits prime consideration. Even in a socialist economy, as in China where enterprises are overwhelmingly state-owned, profit canbe the most important factor in deciding priorities for technology imports. In order to provide incentive to entrepreneurs, managers as well as workers, short-term profit is also emphasized. This is also necessary because of the prevailing shortage of financial capital. Thus, "less investment, quicker economic results" has become a slogan for economic management in China. Indeed, the slogan translates to a higher rate of profit.

Priorities for technology imports are evaluated by certain guidelines:

(1) Readiness of technology for immediate use in the production process;

(2) Ease of access to market, foreign as well as domestic;

(3) Requirement of the "know-how" and of the related preparation for applying the technology to production without interruption.

Understandably, there may emerge a situation which may fail to conform with the long-run economic goals. A case in point is a recent study in Guangdong Province, a province where the share of foreign investment accounts for about 50% of total foreign investment in China (NSBC,1986:582). A study in 1986 points out that as much as 72% of the foreign investment in the province goes to nonproductive enterprises, largely consisting of hotels and restaurants, only 28% goes to productive investments. Similar situations are reported to be found in other provinces and municipalities, including Tianjin, and important coastal "open" city, which has set many successful examples of efficient utilization of foreign investment and technology transfer.

In Hubei Province, the preliminary analysis of a recent survey of the absorptive capacity of imported technology again points to uneven results. To summarize the findings:

(1) Too many small-size projects of relatively simple technology with investment less than US $1 million, and too few large-size projects of larger investment involving advanced technology;

(2) The share of hardware technology is too large relative to software technology, creating bottlenecks for China's technological advancement at a lower unit cost of production over a longer time horizon;

(3) Many duplications of technology imported for the production of consumer goods, which serve only to boost consumption, and thus add to aggregate demand pressure.

On the whole, China appears to have put itself at the lower end of the technology spectrum in order to achieve fast economic results. Shanghai, the most advanced industrial city in China, has been able to do better in terms of absorption of relatively advanced technology, but no firm evaluation is yet available.

There exists a time-cost trade-off in planning an economically sound technology transfer program. From the long-run point of view, China needs to import sophisticated "high" technologies (involving microelectronics, bio-engineering, and basic science), especially the ones for developing energy, transportation, and telecommunication industries, basic to the country's modernization programs. Reliance on this technology transfer path will involve waiting for a longer time in high risk investments before gainful economic results will materialize for China. In addition, access to this path is constrained by a genuine financial burden of very large magnitude.

Technology transfer projects involving smaller financial investments, yielding short-term economic end results may, therefore, be a pragmatic option for China. As the economy progresses on this moderate technology growth path and accumulates surplus savings in the process, the long-run economic development plan may become financially feasible.

China's present policy cannot be described as shortsighted. Ever since 1978, increasingly larger budget allocations have been made for educational and scientific research, and an increasing number of scientific and technical workers, including students, have gone abroad for further advanced training (see Table 3). It should however be noted that deficiency in macroeconomic planning and lack of management expertise continue to limit realization of the full potential of China's scientific and technical manpower reserve. The distorted wage-price system further hurts the system, and a reverse technology transfer, both in terms of brain drain and capital drain, occasionally occurs, as in 1984 (Chang & Lin, 1987; Heller, 1985).

Table 3. Expenditure on Culture, Education, Public Health, and Scientific Undertakings as a Percentage of the Annual Financial Outlay

Year	1952	1957	1978	1980	1984	1985
Percentage	7.7	9.1	10.1	12.9	17.0	17.2

Source: NSBC(1986:32).

Indeed, the critical questions are: how best to reconcile the short-run and long-run economic goals; how to provide enough incentive to investors and entrepreneurs, by keeping the foreign exchange account in balance, so that a difficult credit crunch can be avoided; how best to do the necessary prepartory work to attract foreign investments and how to acquire management expertise to make a system based on imported technology work successfully. China has a great deal to learn from the experiences of other countries, developing as well as developed.

4. Technology Transfer and Product Quality

China, and many other nations in the world today, look Japan in this regard. Japan's superb economic success has been crystallized in its product quality. The country's striving for high quality products not only brings it a trade surplus resulting in a great surplus of financial capital, but also motivates the Japanese entrepreneurs to continue to upgrade production technologies to an even higher competitive level.

In China, quality of products in general is poor, dull in variety and design. For the past many years, except for the intervening years of the "Cultural Revolution", persistent efforts have been made to improve the quality of products in almost every industry in China. It is common knowledge that Japan learnt statistical quality control from Edward Deming and J. M. Juran, among others, of the United States. China should now learn Total Quality Control (a broader and improved version of statistical quality control) from Japan. For years, the Chinese mathematical statisticians (including the present author) have advocated that, as evidenced by Japan's experience, statistical methods such as control charts and orthogonal designs of industrial experiments, plus some operation research techniques, can contribute greatly to the control of product quality in China's industries. It is to be noted that mathematical and statisitcal methods as software technology are relatively less costly.

One reads a news item that the selling prices in Hong Kong are as a follows:

Table 4. Product Origin and Prices
(In HK Dollars)

Product Item	Product Origin			
	Jiangsu Province	Taiwan	Hong Kong	Japan
45-count mixed cotton yarn	2900	3600	3900	—
Grey cloth per yard	2.9	3.62	4.05	6.38

Source: Du(1984).

Table 5. Inter-Country Comparisons of Sectoral Ratios of Fixed Capital to Net Output

China 1981	India 1978—1979	S. Korea 1968	Japan 1965	United Kingdom 1970	Typical Large Coutries	
2.42	4.73	2.43	5.19	6.20	7.69[a]	6.25[b]

Notes: [a] Typical large countries per capita income at US$300(1981 dollars approx.).
[b] Typical large countries per capita income at US$1,200(1981 dollars approx.).
Source: The World Bank (1985:38).

One wonders how prices of the same product, though of different origins, could be so different. Should the difference be attributed wholly and fully to the quality of the product? Or, should there be other factors for explanation, such as trademark, goodwill, financial power, or the power of marketing? As a matter of fact, software technology goes much beyond Quality Control and Experimental Design and Operation Research (which is only a minor part of R&D, which again is a part of the technological infrastructure as it is understood now). It is correct to state, "Only when combined with plant and equipment and with manufacturing, marketing, and financial capacities does R&D result in a commercially meaningful new product or process"(Sahal,1982:32).

Even for R&D alone, statistics point to inadequate appropriation of resources for R&D in China(see Table 5). According to a rough estimate, as of 1983, expenditure on R&D in the developed countries was about 2%~4% of national income, whereas the corresponding figure was only 0.6% in China. Countries like Japan and South Korea allocate a particularly high percentage of national income to R&D (CASS,1985:43; see also Table 5).

Japan's success in producing highly competitive products has been attributed to its extensive technological infrastructure including managerial expertise, a worldwide financial network, captial intensive equipment, a well-trained workforce, and remarkable R&D facilities for absorption of foreign technologies. By comparison, China has paid least attention to international financial networking. Opening of a financial market including the stock exchange is of very recent origin. While reforms in the country's financial market structure is in progress, and while diversified credit services are being explored, how to relate the financial market to the central financial planning in China's socialist economy is still a threoretical problem to be resolved.

5. Free Trade Regime and Technology Transfer

Foriegners have variously described their experiences in trading with China as "the China Fever" turned a sober realism, optimism with caution, two steps up, one step down, depending on the political atmosphere which is volatile. The basic tone, however, is one of optimism.

In China, while trade and technology transfer are being planned and projects are being developed, the infrastructure problem and energy shortage constitute serious bottlenecks. Further, bureaucratic wrangling and structural deficiencies in the industrial system in general act to slow down the progress. But the general trend is one toward problem-solving.

Efforts have been progressively made to reconcile divergent views, and remove existing trade barriers. The move is definitely toward a free trade regime, which will facilitate attracting foreign investment and technology transfer along with it.

Detailed regulations have been enforced by governments at various levels to protect the interest of foreign investors so that the inflow of such investments grow. Provisions were recently promulgated providing favorable treatment for enterprises with foreign investment (Government of China, 1986). Some specific terms of the promulgation are:

(1) Reduced fee for labor employment and land use;

(2) Reduction of taxes, and/or exemption from taxes;

(3) Incentives for reinvesting profits earned;

(4) Guarantees for external conditions required for production priority in terms of access to supplies of water, electricity, and transportation services, communication facilities, for no higher charges;

(5) Granting of autonomy to enterprise-level management;

(6) Protection against local government level arbitrary changes;

(7) Provision for repatriation in foreign exchange of distributed profits;

(8) Expediting contract examination and approval processes.

In addition, specific preferential treatment is given to designated "export enterprises" and "technologically-advanced enterprises".

Autonomy granted to foreign enterprises with foreign investment, export enterprises, and technologically-advanced enterprises deserve further review. It includes the right to determine production and operation plans, to raise and use funds so raised, to determine the wage rates within the scope of their respective contract agreements, and to recruit, employ, and dismiss staff and workers. Thus, these investors will have a much freer atuonomy to conduct their respective businesses in China and make optimally profitable business decisions. This new arrangement is indeed a challenge to China, and China is ready to accept the challenge.

Dissenters may argue that foreign investors are primarily interested in exploiting China's market for their short-term profit, and their concern for China's "export enterprises" or "advanced technology enterprises" are open to question. There is no denying that the balancing of the trade account will be a problem of concern to both the debtor country, in this case China, and the creditor countries, who will have made investments in China.

It simply contradicts the principle of basic welfare economics that a debt-ridden country, such as Brazil, is fighting incessantly for the shortage of foreign exchange, while another country, Japan, is worrying about the optimum disposal of its huge surplus foreign exchange. Therefore, China's efforts in creating a favorable environment for promoting foreign trade and foreign investment warrant careful economic appraisal.

It is instructive to draw upon the following observation (McLean, 1985:239-240).

> Political change will, of course, occur in China, just as it does in any country. However, since returning in 1977 to preeminence in the Chinese leadership, Deng Xiaoping and his closest associates have concentrated their efforts on increasingly institutionalizing reforms that should enable China's long-term adherence to policies of modernization. A great deal has been achieved in the last five years. There is widespread commitment within China to modernization. Increasingly a second generation of leaders and managers have taken on full responsibility for implementing the modernization plans. From all the indicators available there is, in brief, little cause to preach doom and gloom ....

Obviously the process is going to be complex and difficult and China's modernization policies are not going to be carried out without occasional difficulties. Some of them will be too overwhelming. "Considerable forbearance" will be required of foreign investors and others engaged in business in China "in order to overcome such difficulties".

6. Cooperation and Competition

Two basic facts should be noted. First, as of the present time, the great bulk of foreign investment in China is provided through Hong Kong and the largest share of all foreign investment in Hong Kong, in turn, originates in the United States(NSBC,1986:581). This has relevance for a fuller and more informative accounting of the total U.S. investment in China. Second, statistics of Sino-American trade and U.S. investment in China continues to be encouraging. In 1986, trade between the two countries rose to $7.336 billion. The United States now ranks third as an important trade partner of China, and first in foreign investment in China in terms of total contracts signed (Guangming Daily,June 15,1987).

True, Japan has been more successful in trade with China, and has outperformed the United States in this regard. Geographical and cultural ties between Japan and China may account for this. Indeed, China has applied Japanese technologies in many of its key industries, including steel, petrochemical, chemical fertilizer, synthetic textile, and consumer electrical appliances.

As China progresses with its modernization programs, she will need more computers, computerized instruments, microcircuits, electronic instruments, recording equipments, semiconductor production equipment, aircrafts, and space technology, covering a whole range of very sophisticated high-tech industries. The United States will be the natural source for all these (U.S. Emabassy in Beijing, 1984). The increasing technology trade between the two countries cannot be unidirectional for a very long time. There must come a point of reciprocity and mutual gain for both countries.

For the present, China is in a primary state of development when it must be in sympathy with the view that the "third world countries need to embrace the full range of technology, from that which is habitual and historical and comfortable, but inadequate, all the way to that which is at the world standard, which allows some parts of the economy to compete at the world level, so that eventually more parts can compete at this level" (Branscomb,1986:64).

7. Conclusion

While the importance of Sino-American binational economic relations is stressed, China's trade policy should be essentially multinational, based on a principle of "pluralization" in developing the country's technology base. "Technology is forged in the crucible of interests and knowledge," as we have been reminded (Nau, 1986; Fusfeld & Haklisch, 1982). While self-interest leads the way to competition, knowledge provides the basis for cooperation. May competition and cooperation coexist in such a way that the future economic development of all nations will follow only the first three stages of the Kindleberger Cycle, namely, imitation, innovation, and invention, but not the fourth, loss of technological prowess to any one individual country.

Note

The New York Times (July 29, 1987) published an article, later translated into Chinese by CANKAO XIAOXI (September 24, 1987)... of the daily output of $1.6 billion of Japan, as much as $0.79 billion goes to the foreign exchange market and related speculative investments. ... The continued favorable balance of trade further strengthens the yen, and this leads to a series of problems: higher rate unemployment, insufficient tax revenue, reduced government expenditure, shrinking domestic demand, and once more recourse to export market. Thus, a vicious circle perpetuates.

Should Finite Population Inference Follow a Different Principle?[①]

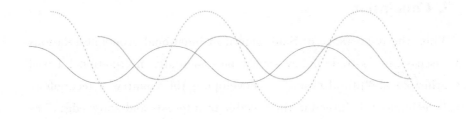

1. The Challange

For the last thirty some years, a number of noticeable theoretical achievements in the area of finite population sampling have contributed to the forming of a school of thought that places a potential challange to the time-honored principles of sampling laid down by Neyman (1934). A "new" framework has been designed by Go-dambe (1955) and others of the Indian school to preach the "unified theory" parallel with the traditional concepts of statistical inference. Theorems and cases were presented to indicate the inadequacy of the conventional survey sampling theory and practice. Someone even expressed his distrust of the conventional concepts with such a passion that "Alongside with the concept of significance test and that of confidence interval, the concept of unbiased estimate is one of the three most widely used, most controversial, and most misleading notions of statistics" (Basu, 1978). A situation as such has, naturally, not merely stimulated interest among theoretical statisticians, but also caused concern among survey practitioners. Review and comment papers (Godambe, 1969; Smith, 1976; Hansen & Madow, 1978) on the foundations of survey sampling have appeared from time to

---

① 原文为在"The second Japan-China Symposium on Statistics"上交流材料。Kyushu University, Fukuoka, Japan, 1986 年 11 月。

time. In this paper, I am attempting to follow this fascinating development of statistical thoughts with a few remarks of my own.

2. Doubts about the MVU Principle and "Therefore"

Inspired by the basic paper of Horvitz and Thompson (H&T)(1952), Godambe (1955) proved that no minimum variance unbiased (MVU) estimator could exist uniformly for all possible populations within the class of linear estimators, which can be expressed in the form

$$\hat{T} = \sum_{i \in s} \beta_{si} Y_i$$

where the values Y_i under study are associated with unit $i(i=1,\cdots,N)$ in the population, and the coefficients β_{si} depend on the unit i and the samples. Note that this formulation of linear estimators presupposes identifiability of the individual units.

The impossibility to derive a MVU linear estimator cast doubts upon the feasibility of Neyman's approach to finite population sampling, and alternative approaches were tried. As the likelihood function was claimed to be "the only bridge that links the observed sample to the unknown parameter", a likelihood function for the finite population, with $\omega=(Y_1,\cdots,Y_N) \in \Omega$ as a universal parameter, was stated (see, e.g., Basu, 1978). Thus, let S be the survey design, $x=(s,y)$ the sample, where $s=(i_1,\cdots,i_n)$ with $i_1<i_2<\cdots<i_n$ is the identifier (label) set, and $y=(y_1,\cdots,y_n)$ with $y_k = yi_k$ is the observation vector. The likelihood function has the following form:

$$L(\omega \mid x, S) = \text{Prob}(x \mid \omega, S) = \begin{cases} q & \text{if } \omega \in \Omega_x \\ 0 & \text{if } \omega \notin \Omega_x \end{cases}$$

where $q=q(x,S)$ depends only on x and the design S. Ω_x is the set of parameter points in Ω that are consistent with the observation x:

$$\Omega_x = \{\omega : \omega \in \Omega \text{ and } Yi_k = Y_k (k=1,\cdots,n)\}$$

It is to be noticed that, given the sample x, one can determine Ω_x without any reference to the design S. Then, it is pertinent to ask: why should we pay so much attention to survey design? In particular, why should we use randomization instead of purposive selection?

Furthermore, given sample x, all parameter points in the set Ω_x have equal likelihood and all points outside Ω_x have zero likelihood. This flatness of the likelihood function, being not informative about how a single parameter point may be selected, may nevertheless offer a good opportunity to conceive some suitable model or prior distribution for predicting the true value of the parameter. It is further argued that some model-based analysis is necessary to relate the observed part to the unobserved in order to make the prediction possible.

So the logical sequence in the development of a "new" theory of finite population inference would be:

Lack of a best unbiased estimator$\Rightarrow$
  utility of the likelihood function$\Rightarrow$
Opportunities for model-based analysis and purposive sampling.

3. Relevance of the Identifier

All these seem fascinating and pretty confusing. But the crux of the problem is the relevance of the identifier peculiar to finite population.

The reason why the sampling plan plays no part in the analysis of the likelihood function when the data are given is implicit in the fact that we want a universal parameter $\omega = (Y_1, \cdots, Y_N)$. But it is extravagant to ask of any statistics to predict individual values associated with individual units in the population. This is not even statistics which is supposed to be concerned with "averages". Should we be willing to do away with some of the information provided by the identifier at the stage of analysis, we will rid ourselves of the perplexities. Suppose we are satisfied with asking how many units in the population having a value in the interval $(x, x+\Delta)$, we will be able to write the likelihood function in a form not merely informative about the parameter, but also dependent on the survey design. As many more intervals are made, each being as fine as we wish, we shall come to a good description of the empirical (sample) distribution of the population under study, and this will bring us enough information that statistics can provide.

Therefore, it is not unreasonable to conclude that though a flat likelihood can be made in singular cases under certain assumptions somehow informative about the universal parameter, the conception and formulation of the universal parameter system should in general be refuted, at least as impracticable.

4. Fundamentals of Neyman's Principle Revisited

In his monumental 1934 paper Neyman said:

"If we are interested in a collective character X of a population and use methods of sampling and of estimation, allowing us to ascribe to every possible sample, Σ, a confidence interval $[X_1(\Sigma), X_2(\Sigma)]$ such that the frequency of errors in the statements

$$X_1(\Sigma) \leqslant X \leqslant X_2(\Sigma)$$

does not exceed the limit $1-\varepsilon$ prescribed in advanced, whatever the unknown properties of the population, I should call the method of sampling representative and the method of estimation consistent."

Thus, representativeness of a sample is defined through the concept of confidence interval in an objective way, and is to be achieved by the powerful central limit theorem in the theoretical aspect and by the largeness of the survey sample in the practical aspect. As a consequence, we can make "predictions" with the observed about the unobserved (inferences from the part to the whole) even if the population comprises axes, asses and boxes of horseshoes (Royall, 1971). This achievement is beautiful and infallible.

It should also be noticed that representative sampling and consistent estimate go hand in hand. Representativeness implies a "stochastic" relation between the observed part and the balance of the population. So, in answering the above question of "why randomization", we suggest that randomization implies representativeness, which allows "predictions" with given "precision".

It is perhaps the efforts to seek a "most" representative sample and a "best" estimator that lead to controversies.

5. A Practical Viewpoint

It is believed that randomization is a protective measure against human bias. Though sometimes mathematical unbiasedness sounds silly, we can not do away with common sense unbiasedness. It is good and even necessary to have a survey result to be demonstrably unbiased. Simple random sampling is easy to be so demonstrated. Even the H&T estimator for unequal probability sampling can be compared to an estimator under simple random sampling in the following simple way: when a unit is selected with probability p, this amounts to simple random sampling after multiplying the number (1) of that unit with a factor proportional to p. Therefore, once that unit is selected, its value should be deflated by the same proportional factor. Normalization of the factor gives the H&T estimator.

However, mathematical unbiasedness means sampling unbiasedness, which is only desirable for large samples rather than small samples. Purposive selection in connection with an assumed model is acceptable for small samples when the model has been demonstrably consistent with the data. Even for large samples, the sample results are often compared to some well established models, say, the log-normal distribution of income in a socialistic society. Of course, model buildings are often complicated by multi-purpose sampling.

While Godambe and others (Basu, Royall, Rao, Ericson, to name a few) have made significant contributions to the understanding of finite population sampling problems, I do not see much connection between the theorems and cases they presented and the sampling practices such as purposive selection as advocated by some of them. For, as a matter of fact, the traditional sampling practice has not missed the chance of using whatever prior information available to aid survey design and analysis including data screening, post-stratification, etc.

I believe that while the art of model building deserves a greater attention, especially in the areas where data are dear or randomization is inconceivable (e. g. , non-response analysis), the effort to seek improved variance estimates of various estimators should grow.

Should finite population inference follow a principle very different from Neyman's? Probably not.

References

Most of the works cited can be found in J. Royal Statist. Soc., A 1976, B 1955, 1966; Symposia: New Developments in Survey Sampling, Wiley, 1969; Foundations of Statistical Inferences, Holt, 1971; Survey Sampling and Measurement, Academic, 1978.

附件：在日本创价大学的演讲
（1986年11月11日）

尊敬的高雄学长、各位教授、女士和先生们：

感谢创价大学对我们的邀请，感谢砂田吉一教授从中做的努力。我是一位数学工作者，又是统计和经济工作者。这次是在福冈九州大学参加中日统计讨论会之后来东京创价大学顺访的。通过第一、二届中日统计讨论会的召开，我们比较充分地认识到了中日统计学术交流的重大价值，不仅促进了双方统计学的发展，互相取长补短，还增进了我们两国人民之间的友谊。我们中方的统计学者感到第二届比第一届开得更好，有了更多的共同语言，并发现有更多的合作领域和题目。例如，砂田教授在第二届会议上提出了关于公司财政（评价公司业务营业）的有价值的论文，我们也有意尽早开展一些跨国公司业务经营方面的研究，还有关于中日两国在抽样调查、品质管理方面的异同等。这里举的只是一些和经济领域有关的课题，当然还有其他方面的课题，就不详列了。

前天下午在中日双方组织委员的联席会议上，大家一致推举砂田教授为下届中日统计讨论会日方组织委员会委员长，并且决定第三届的地址为东京创价大学，我们想到创价大学在日本独特的、崇高的地位以及砂田教授在经济学和数量经济学界的声誉及活力，都感到非常高兴。在统计学用于经济振兴和发展方面，我们有很多东西需要向日方，包括向砂田教授所在的创价大学经济学部学习。

这次到日本来，我个人已经有了初步的许多美好印象。前天我有机会和浅野（Asano，前任委员长）进行较长时间的交谈。他幽默地说："日本没有长城、故宫等让你们参观",但我们在日本看到许多现代化的东西,不仅是物质文明,还有精神文明。我们亲眼看到科学技术怎样变成生产力,严明的管理制度和有效的在职教育等。

因此,我想借此机会谈一点当前在中国开展得相当广泛的一个课题,以便向日本朋友和专家学者请教。

如何衡量科技进步在国民经济增长中的作用。我和我的同事,在国家经济委员会的倡议和支持下,已经进行了三年的研究,我们发现许多地方需要借鉴于外国,当然社会主义经济是以马克思主义经济理论作为指导思想的。我们首先遇到的一个问题是常常用来计算技术进步的 Cobb-Douglas 生产函数是否适合于社会主义国家。随之而来问题就是如何估计其中的参数,笼统地说有所谓份额法,但是我们不接受"工资率=劳动边际生产力,利息率=资本边际生产力"的利润最大化原则。然而,我们的一些研究工作者仍然主张用份额法,根据产品的价值 P 分为固定成本 C、可变成本 V 和剩余价值 M 三个部分,就取资本份额为 $\alpha = C/P$,劳动份额为 $\beta = (V+M)/P$,这样就出现了一个国际可比性问题。

另一个估计方法是回归。如果不作规模报酬不变,即 $\alpha + \beta = 1$ 的假定,计算的经验往往出现负的 α、β 值,出现负值的频率竟然高达 1/4 至 1/3。投入增加,产出反而减少,当然不合理,说明数据有问题或管理不善,或者是所谓多重共线性。当我们假定 $\alpha + \beta = 1$,就基本上避免了负值的出现,但是我们还总是认为管理效率是一大问题。在国外文献中也偶尔出现负的 α 或 β 值,有人解释为简单的回归,忽略了价格。在中国,价格不合理我们已经公开宣布了,正在进行改革。

为了显示管理效率方面的问题,我的一位研究生作了前沿生产函数的估计,由于缺乏可靠数据,结果虽然粗略,却引起广泛的兴趣。

日本在 C-D 函数的参数估计,我们也作了参考和对比。我们的研究当然不限于 C-D 函数的应用,我们对结构变化如何改变 K、L 在各 Ind. 之间的分配也作了评价,时间所限就不详细介绍了(见论文集)。中国很重视这一研究,什么最能促进经济增长,很多数学、统计和经济工作者参与这项工作。

最后有两个建议:(1)我代表华中工学院经济管理学院希望今后能和创价大学经济学部建立某种学术交流;(2)我代表中国现场统计研究会,建议通过下届日方组织委员会的联系,和日方 QC 方面的经济或统计工作者在 1987 和 1988 年分别举行一些小型专业讨论会,我们愿意在 1987 或

1988年先在北京接待日方的 QC 专业工作者。

再一次感谢创价大学和砂田教授给予的热情和周到的安排。

（根据林少宫教授的原始手稿整理）

［后记］1986 年 11 月 11 日，参加日本福冈九州大学举办的第二届中日统计讨论会后，华中工学院经济管理学院院长林少宫教授和北京大学前校长张龙翔教授等应日本创价大学的邀请访问了在东京郊外的创价大学并和该校创办人池田大作先生会晤。在校园，他们受到师生代表们的夹道欢迎，瞻仰了"周樱"（纪念周恩来总理的樱花树，蕴藏着池田与周总理的深厚友谊）。访问期间，林少宫教授作了技术进步测度的演讲，张龙翔教授作了中国教育改革的演讲。

China's Modernization: Stability, Efficiency, and the Price Mechanism[1]

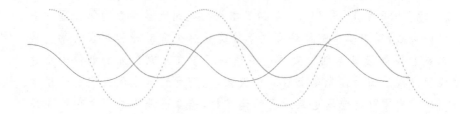

1. Stability, a Priority

"Reform of China's economic structure... should serve to advance, and not to impair, social stability, expansion of production, improvement of the people's living standards and the growth of state revenue."—This citation from a "Decision" taken by the Central Committee of the Communist Party (1984) may provide a key to understanding China's economic policy. Over the years, China's economic policies have been formulated not merely in the interests of social stability, production growth, and the people's livelihood, but also, at least in retrospect, in this order of priority.

There have been virtually no exceptions to the above rule when China's price policies were being formulated, as revealed by the historical data:

(1) During the years 1950-1983, the retail price index in China went up only 55.6%, averaging an annual increase of 1.35%, whereas over the same period the prices of consumer goods increased variously some three-to ninefold in West Germany, the United States, Japan, France, and the United Kingdom (*Almanac of Statistics—China*, 1984). China's

---

[1] Asia Pacific Economies: Promises and Challenges. Research in International Business and Management, Volume 6, Part A, pages 103-118. Authors: Chang Peikang and Lin Shaokung.

comparatively stable price level contributed a great deal to the stability of its economy.

(2) Beginning in 1957, the rate of exchange between the manufactured goods bought and the agricultural products sold by the peasants has been readjusted from time to time in favor of agriculture so as to diminish the so-called scissor differential, an expression of exchange of unequal values between agriculture and industry. Thus, in 1957, the rate of exchange was restored to about the level of 1937. Again, in 1983, the purchasing price of agricultural produce was raised to 321.3 as compared with 100 in 1950, while the selling price of manufactured goods in the countryside went up only 14.8% and that of means of production for agriculture increased by only 8.4%. As a result, the peasant could trade, in 1983, with only 35.7% as much of the agricultural produce as in 1950 for the same amount of manufactured goods (Editorial, 1984). This narrowing of the scissor differential is generally considered an important step toward maintaining and strengthening good relations between the workers and the peasants.

(3) Ever since 1953, the year when price controls were initiated in China, the prices for minerals, raw materials, and energy, which are considered basic to industrial production, have been kept (artificially) low. This has had a stabilizing effect on the general price level in as much as fuels and raw and semifinished goods constitute a sizable part of the costs common to many factories.

(4) Throughout the years since the founding of the People's Republic of China, prices for daily necessities have been kept low and stable to ensure that sufficient means of livelihood are accessible to the people. Grain, edible oil, heating coal, etc. are rationed at low prices. Housing, medical care, newspapers, etc. are provided either at nominal prices or free of charge. This is the salient phase of the socialist economy, which lays the foundation for achieving the overall economic stability in society.

The above points cover the various aspects of China's price system that have brought stability considerations to the fore. It seems that China has acted as a grand risk-averter in the price domain, and indeed made great achievement as far as social stability and security are concerned. However, in this paper we wish to devote ourselves more to the problem of economic efficiency in connection with China's price system and examine what should be done when both stability and efficiency are deemed important.

2. A Retreat From Efficiency

There is no denying that stability is always a factor that carries great weight in the mind of any policymaker. But by what means is stability attained, and to what extent? In China, practically not a single enterprise has ever gone bankrupt, and not a single household need seriously worry about its regular income. This situation has been described as "everybody eating from the public big pot." This is something more than stability. It is simply safety! But has this anything to do with the price system? Yes, indeed. This is largely a result of the system of rationing and price control.

In China, the prices for mineral products, raw materials, and energy—coal in particular—are held extremely low, perhaps the lowest in the world. In fact, they are *far too low*. However, many products turned out by our processing industries with such cheap raw materials and fuels cost far more than comparable goods sold in the international market. To cite an instance, the revenue from exporting one ton of granular active charcoal is \$800, whereas exporting the fuel necessary to extract that one ton can bring about \$1680. Therefore, it is only artificial to report the profitability of those fuel-consuming enterprises, as the fuels are artificially cheap. Because of the rather confused price relationships, it is virtually impossible to properly evaluate the performance of enterprises in general.

Our cheap raw materials and fuels do not provide the processing industries with better conditions for international competition as they should, but rather discourage efforts to seek technological and managerial improvement. By way of further illustration, the price ratio between crude oil and gasoline in the 1980s is 1 : 1.24 in the international market whereas it is 1 : 7.2 in China. It is estimated that the state suffers a loss of about 1 trillion yuan due to low energy prices maintained over the past 30 years (Chinese Energy Economics Research Group, 1983).

It is noticeable that the retail prices of major agricultural and ancillary (i.e., "sideline") products are lower than their state purchasing prices. This is a phenomenon resulting from stability considerations. For, on the one hand, we want to keep the cost of living low by leaving the retail prices on grain, edible oil, pork, eggs, etc. essentially unchanged and, on the other hand, to raise the state purchasing price for agricultural and sideline products so as to gradually bridge the huge gap between the prices of industrial and of agricultural products—the so-called scissor differential inherited from history. The effort to seek social stability has brought about the abnormal phenomenon of buying high and selling low. Such an inverted relationship between the buying and selling prices poses a fundamental problem. Is it justified for the state to subsidize the retailer for selling low rather than to pay a higher wage to the wage earners for buying high? It is not just the form of payment but the freedom of choice on the part of both the consumer and the producer that matters. It is also a part of the general economic problem of how much should be planned (planned production and guided consumption) and how much should be left to the market.

The practice of price subsidization that began with a single item, cotton for wadding, in 1953, was then extended to 5 items in 1960, 11 items in 1970, and 38 items in 1980. At present, among the subsidized items are the following: grain, edible oil, pork, aquatic products, poultry, eggs, beef, vegetables, cotton for wadding, silk and silkworms, cowhide, diesel oil and electricity for farming, electric bulbs, pig iron pans, textbooks, exercise books, newsprint, small farm tools and appliances, farm machinery, etc. It is estimated that about 35 billion yuan are spent for price subsidies each year, including subsidization of the management cost involved in dealing with subsidized commodities and house rent. Of this amount, 70% goes to subsidize daily necessities, 10% to support farm production, and 20% to underwrite the loss entailed in dealing with imported goods (Qiao, 1983). There have been a number of disadvantages associated with price subsidies in addition to the state financial burden, namely, (Ⅰ) weakening the function of price as an economic lever; (Ⅱ) deterring the improvement of management; (Ⅲ) restricting the production and circulation of subsidized goods; (Ⅳ) inducing a rise of the general price level—a consequence of the relative decrease of public expenditures for accumulation; and so on.

In China, it is not unusual that products of different quality are not differently, or at least not sufficiently differently, priced. Entrepreneurs are not encouraged to collect the "rent". As a result, poor quality goods drive out fine ones. While the consumers waste their time in queuing up for the scarce fine-quality goods, most producers are content with turning out the same old-fashioned products year after year without ever a thought of updating them even when they pile up in stock. Similarly, price differences between different varieties and for different localities are not administered as they should be. As an obvious example demonstrating the consequence of the rigidity of price without due regard to varieties, there were, in the early 1950s, hundreds of varieties of dried and fresh fruits sold in the market in many cities, but nowadays only tens of them remain owing to the unfair practice of fixing one and the same price for different grades or varieties of goods.

Failure to sufficiently differentiate prices (say, for coal) between different localities has a perverse effect on rational allocation of industries.

True, it is impractical to ask the government to adequately fix a price for every variety (goods of different qualities; goods in different locations treated as varieties). Rather, decentralized management is needed and market mechanism can be utilized.

Complaints and criticisms are often heard about poor services in China, and, efforts have been made to improve the situation. However, service industries have developed slowly due to low charges. There are traditional and ideological reasons for this. For years, commerce and commercial services have been looked down upon and neglected. The prevailing ideology is that service industries do not contribute to national income and that it is not morally justified to charge a fee for merely rendering a service. Such ideas are certainly incompatible with modernization requirements. The lack of adequate service has caused numerous inconveniences to the people's livelihood and deterred the development of a modern economy. In particular, an adequate information service has relevance to improving the economic return in domestic and foreign trade. Admittedly, some services are difficult to evaluate and only international standards are available for reference. As such, they can hardly be put under price control. Perhaps the service market can tell how much a service is worth.

In the final analysis, stability through excessive and rigid price control is itself debilitating. On the one hand, enterprises do not have to struggle to survive; on the other hand, people have to be content with their status quo for lack of work incentive. Industries providing services of various kinds have not developed as fast as they should have. Where price fails to properly function as an economic lever, inefficiency results from absence of an adequate measure of efficiency.

3. Formation of Price in a Socialist Economy

For each commodity to be sold and bought in the market there must be a price. But not everything is a commodity. Goods can be allotted and delivered to the producer or the consumer without a price or only at a nominal price. In the early years of socialism, commodity production is limited to individual consumer goods. Means of production are not considered as commodities and, therefore, not to be sold and bought in the market but allotted to the producer through state planning. As a result, many goods are not properly priced so as to balance supply and demand; some goods get stockpiled while others fall short in supply. Price, the money value of a commodity, does not function as it should. For lack of commodity production, the economy may be seriously distorted and efficiency lost.

Now, at least in China, it is explicitly alleged that the socialist planned economy is a planned commodity economy based on public ownership (of means of production). Not merely consumer goods but also means of production belong to the category of commodity. Although means of production are public owned, they are put under separate managements, so the commodity relation still applies. Furthermore, since the consumption of means of production is ultimately related to the production of consumer goods, it is simply illogical to identify one but not the other as commodity. The traditional idea of pitting the planned economy against the commodity economy must be discarded.

Owing to China's rather undeveloped commodity production at the present stage, a full development of the commodity economy is indispensable for the socialist economic development, and indeed a prerequisite for her modernization program. But it should be noted also that the extensive growth of a socialist commodity economy may lead to certain disorder in production due to market competition, a necessary companion of commodity production. Hence the need arises for a planned commodity economy, which combines the planned economy with the commodity one. This is possible because the law of value, as will be explained shortly, that characterizes the commodity economy will be consciously followed and applied in a socialist planned economy. In order to act in conformity with the law of value, on the one hand, the state needs market information in setting its plan and, on the other hand, individual producers need guidance from the state to avoid chaos in production. Thus, a commodity economy is required independent of the social system. The difference between socialist and capitalist economies, as far as commodity economy is concerned, lies primarily in the difference in ownership, in the different purposes of production, and in whether the law of value can be consciously applied. In addition, the scope of commodity relations are different. Under China's socialist conditions, neither labor power nor land nor mines, banks, railways, and all other state-owned enterprises and resources are commodities.

In the socialist economy the production and exchange of commodities should be governed by the law of value. According to Marx's theory, the value of a commodity is determined by the socially necessary labor time required for its production, which implies that (1) not merely the socially necessary labor time spent in producing one unit of the commodity (2) but also the necessary allotment of the total labor time available in society to the production of the kind of commodity in question. This means that only an amount of the commodity as demanded by society as a whole should be produced. Thus, the law of value presumes an efficient allocation of resources when the production of a commodity is interpreted as a labor process. Socially necessary labor time is a statistical concept. It is the labor time averaged not only over different units of production but also over all the producers in society.

Price is the monetary form of value. Given technology, the value of a commodity may be represented by a horizontal line indicating the socially necessary labor time embodied in the commodity, and so is the price, P, when the law of value is effectively applied. But the quantity, Q, produced is determined by social aggregate demand (see Figure 1). Those who produce with labor time less than socially necessary can make a profit higher than normal, and those with labor time more than socially necessary will attain zero or less than normal profit. In a planned economy, information is needed in order to get to point Q. Actually a process of adjustment and readjustment is involved short of perfect information. To alleviate shortage or surplus in the short run, price may be allowed to fluctuate within controlled limits with value as its center. This practice only means a planning system that combines uniformity with flexibility. In the long run, shortage and surplus will cancel out in the process of conscientious adjustment of production.

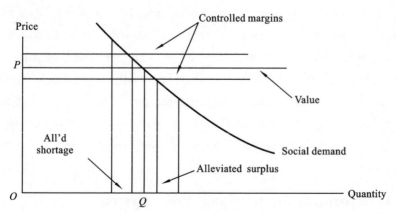

Figure 1

It is worth pointing out that the shift of demand should bring about only a change of the quantity produced but not the price, unless there is an economy or diseconomy of scale. An enterprise should never try to increase its income by a price increase. Only improved qualities can bring about better economic results. It is expected that the better the quality of a product, the higher the demand for it (shift of demand), and a higher price is justified. Thus, the quality-price (or demand-price) curve is one shifting upward to the right.

Obviously, statistical information on socially necessary labor time required for producing each commodity is necessary for successful management of the planned commodity economy. Data from the market channel should be carefully observed, collected, and utilized.

The development of commodity production necessarily leads to quality-price competition among fellow producers. Enterprises are to be put to the test of direct judgment by consumers in the market so that only the best, i. e., most efficient, ones survive. However, as conscious applications of the law of value imply exchange of equal values, socialist competition is aimed not so much at increase of profit as at increase of efficiency. Cooperation and association naturally take place along with competition.

It is important to observe that the application of the law of value is consistent with the principle of distribution according to work. Suppose every producer spends just the amount of socially necessary labor time in doing the same work. It is plain that the reward will be commensurate with the work done when the law of value is applied. But if anyone who is more efficient need spend less time than is socially necessary in doing the same work, he or she should get the same pay. This means that if that person were to spend the same amount of time on the job as thought socially necessary, he/she could do more work and therefore should get higher pay.

4. Reform of the Irrational Price System

The law of value provides a norm for price reform. However, not every departure of price from value should be corrected or removed right away. Individual departure may be desirable from the viewpoint of national security or long-term development. For example, some staple farm product may continue to be selling low and so require a subsidy to be provided for a long time to come. Yet, apart from normal price fluctuations, departures of price from value are, generally speaking, sources of trouble rather than desiderata. An investigation should be made with respect to their causes and remedial measures sought accordingly.

The irrational system of pricing is inherent in the practice of excessive and rigid price control. Steps should be taken to gradually reduce the scope of uniform prices set by the state and appropriately enlarge the scope both of floating prices(to be negotiated within certain limits)and of free prices. For example, it is advisable to adopt a flexible price policy for some small commodities and farm and sideline products of the third category to meet market demand, and also for fresh and live products and products marketed after state purchase and delivery quotas have been fulfilled.

Thus, the State Planning Commission has decided, beginning in 1985, to narrow the scope of mandatory planning and extend that of guidance planning and market regulation by cutting the number of industrial products of a mandatory nature from 120 to 60, and the number of agricultural and sideline products to be purchased by the state from 29 to 10. This decision will grant enterprises more decision-making power and spark the producers' enthusiasm and initiative to run the businesses better. It is expected that prices will then respond more quickly to changes in labor productivity and to shifts of market supply and demand, and thus better serve the needs of the economy.

Meanwhile, the overcentralized price control system needs some degree of decentralization. Because of lack of adequate information about the cost and labor productivities for a multitude of minor commodities, it is impossible for the State Price Commission to set a right price for each commodity. Various price-administering powers should be delegated to lower-level authorities to suit local conditions.

Reform of the present price system is in essence a readjustment of relative prices. For historical and other reasons, there have been many commodities the prices of which reflect neither their value nor the relation of supply to demand. For these commodities the cost-price relationships are so distorted that no correct assessment of the performance of the enterprises concerned is possible. For some commodities (say, some raw materials) there seems to be a chronic shortage just because the relative prices have become far too distorted. For others (say, some mechanical products) there may be a surplus, only concealed by state purchase. Therefore, it is advisable (I) to raise the prices for some raw and semifinished materials in anticipation of a new production mix, and (II) to lower the prices for high-grade consumer goods along with an increase in supply.

Price reform should be aimed at readjusting structural prices so as to create a favorable condition for competition between enterprises. Profitability of an enterprise must be based on its performance, and not on price artificialities. Only when man-made price differences or price uniformities are essentially removed will competition among enterprises lead to genuine economic efficiency.

Under a system of rationing and rigid price control, the state-owned enterprises are run as if wage cost and price were unrelated. In fact, the price of a commodity cannot be set right without due regard to wages, taxes, subsidies, sales expenses, profit, bonus, etc. In the same way that the price of a commodity is related to its value, so the wage and bonus paid to a worker should be related to his/her performance, and so profit or after-tax profit should be related to the managerial accomplishments of the enterprise. The reform of the price system would not be complete without a reform of the systems of wages, taxes, subsidies, and the like.

A few years ago, a system of contracted responsibility was initiated in the rural area with great success in promoting production. It was a system of assuming economic responsibility by individual households for contracted jobs, under which farm products could be owned by individual households provided that they delivered their quota of grain to the state, paid off their loans, and allocated sufficient accumulation funds and public welfare funds to the production team. The basic experience of the responsibility system is applicable in various forms to the urban enterprises because it conforms with implementation of the law of value and the socialist principle of distribution.

Thus, as a practical step to implement the principle of distribution according to work in urban enterprises, the long-practiced system of profit delivery in China has generally been replaced by a new tax system since 1983. It is the first time in more than 30 years that the economic relationship between the state and enterprises has been fixed in the form of taxes. State-owned enterprises which used to hand in all their profits to the state now pay taxes and keep the after-tax profits. This is an important step to better link income with performance. Indeed, this is a bold step! It is estimated that the state treasury is facing an immediate loss of 3.5 billion yuan a year due to the new tax system, but it is expecting a long-term return when enterprises use the after-tax money to expand production.

Also, since 1983, a policy has been formulated to the effect that enterprises may decide on the amount of bonus to be paid to their workers and staff members according to the results of enterprise operations. The state only collects an appropriate amount of tax on the above-norm bonus from enterprises.

In essence, we are developing and putting into effect a system of work incentives that best suits China's existing conditions.

Judging from China's present wage system, the difference between the wages of various trades and jobs should be further widened. Payments should more fully reflect the differences between intellectual and manual, complex and simple, skilled and unskilled, and heavy and light work. In particular, attention must be paid to the fact that the present remuneration for intellectual work is relatively low and should be appropriately raised. The misconception of socialism as egalitarianism is utterly incompatible with the scientific Marxist view of socialism.

5. Avoiding Inflation and Ensuring Success

Now China has openly announced that the reform of her price system is under way. It is pertinent to ask: What is the prospect for success? Will it cause inflation? The answers depend on what practical steps are to be taken and on what conditions are provided and to be created.

Because the present price system is seriously flawed and there are many shortcomings to overcome, the reform must not be carried out in haste. It is prudent to start the reform with the easier tasks in *small* steps, as is now planned. For example, the readjustment of purchasing prices and the price ratios between farm products of different varieties and qualities so as to achieve a more coodinated agricultural production structure can be done first. Also, the readjustments of prices for some primary industrial products and capital goods that have little direct bearing on the people's livelihood are also among the easier tasks. Such small steps may be noted in the limited range in which prices are allowed to vary. Even if the readjusted prices, as judged from some optimality criterion, are still far from right, no abrupt alterations will be made. The readjustment of prices for products like grain, edible oil, and rent which have a great bearing on the people's livelihood may be postponed for the time being. It is planned that the present price reform shall take about three to five years to carry out.

It is worth noting that in China people have been used to a system of rigid prices. Even if small steps were taken on price increases, a rash of buying might occur. The psychic effect and chain reaction following a change in price must not be slighted. As a matter of fact, in this regard some past experiences may be drawn upon. For example, beginning in 1979, the prices for grain, edible oil, and pork, items of importance to the people's livelihood, have been raised by small margins together with corresponding subsidies provided for the consumer. This was considered a small step. However, it has led to a repercussion of price rises on numerous consumer goods and daily items, so numerous as to defy any effective market inspection and control! Meanwhile, the bonus system, originally initiated as a means of work incentive, was widely adopted and played a negative role in the vicious high price-high bonus circle (or, rather, spiral). We actually faced the threat of a mild inflation in 1980 and again in 1984 (see the Appendix). However, thanks to the state's regulatory power over the market, the tendency toward rising prices was checked. Admittedly, too much exercise of state regulatory power may cause a short supply of some commodities. However, no serious problems arise as long as the discipline of production is maintained and purchasing power is no further expanded but, rather, is adequately curbed. It is characteristic of the socialist planned commodity economy that inflation can easily be checked, though perhaps at the expense of a temporary (and sometimes prolonged) shortage of some commodities. A relevant experience from the past was that of the price stabilization period following the Great Leap Forward Movement in 1958 (see the Appendix). But this time we must keep an eye on the possible detrimental effect of price restoration on the development of production!

The readjustment of prices for raw materials, crude oil, coal, and mineral products is also a tough task. However, it may not produce an effect as serious as at first appears. It is expected that the substitution effect will lead to economical use of those goods whose relative prices rise. As the prices for raw materials and fuels increase, new technology will be introduced and a change in production mix will take place. Consequently, the cost of production will not increase as much as was calculated in the absence of technological changes. Furthermore, socialist enterprises may take advantages of the fact that the state is always ready to provide necessary subsidies, at least for a short period of time, and technological guidance to overcome the difficulty created by the change of prices.

Similarly, a higher relative price set for some staple farm products not only changes the agricultural production structure but also affects the consumption pattern, including people's diet. For example, for the last two or three years, the consumption of rice has been greatly reduced as more meat, eggs, confectionery, and other higher-grade foods have become available. This is a combination of income and substitution effects. Inferior foods have given way to higher-grade ones. Thus, the ease and smoothness in carrying out the price reform depend in a sense on the adaptation of new technology and the abundance of substitutes.

It is peculiar to the socialist economy that in the process of price reform there is no secret as to what prices are to be raised and what to be lowered, and at about what time these will be put into effect. It might seem that such an advance notice of price changes might provide a chance for enormous profiteering by speculators. However, we are confident that people will show their cooperation in the price reform, given proper education on the price policy. The announcement of wage increases may also have a psychic effect on the general price level, but the undesirable consequences will be minimized. The state authorities have repeatedly made it clear (I) that a calculated plentiful stock of commodities are in reserve to cope with the potential increase of purchasing power resulting from wage increases, and (II) that wage increases will not overshoot those of prices, although the real income of wage earners will keep on gradually rising.

Education on price policy, besides price decrees and regulations, is always an important factor in China that provides additional safety margin for the price reform.

Of course, the more important and real things to guard in the battle against inflation are the physical conditions provided. Based on past experiences, the greatest danger comes from the overexpansion of capital construction. Therefore, conscientious control must be exercised over the scale of capital construction. A basic balance between finance, credit, and materials must be struck to coordinate the effort for price reform. The circulation and issuance of currency must be kept under strict control. According to the Seventh Five-Year Plan (1986—1990), efforts will be concentrated on the construction of key energy and communications projects; capital construction investment will be focused on the transformation of old enterprises. Construction of other projects must also be kept under strict control.

Now, the supply of grain, cotton, edible oil, and other farm products is abundant following the recent bumper harvests in recent successive years. An ever-increasing number of designs and varieties of goods are being turned out and sold in the market in consequence of the rapid growth of agricultural and industrial production and commercial activities. The supply of and the demand for most commodities are basically balanced. Goods—some consumer durables, for example—that happen to be in short supply can be imported when necessary. Economic opportunities provided through the use of foreign capital and technology are enhanced by increasingly pursuing an "open" policy. The settlement of the Hong Kong issue by means of "one country, two systems" and the enlargement of the special economic zone play no small part in promoting foreign and domestic trade and invigorating the economy. Currently, efforts are being made with success to curb price rises as induced by the excessive issuance of currency and the wide distribution of bonuses in 1984. It is thus believed that the conditions for reforming the price system have ripened.

To sum up, while relative prices must be changed, the general price level should be kept as stable as possible. China's price policies have been successful in keeping a relatively stable price level throughout the years except for the period of the Great Leap Forward beginning in 1958. However, the price structure is seriously flawed due to excessive and rigid price control. A thorough reform aiming at readjusting relative prices is now in place, but the steps should be *small* and *gradual* so as to minimize undesirable repercussions. As production keeps on expanding, a mildly rising general price level is likely to occur (and is perhaps desirable in some sense). But when the growth of the economy is kept at the right pace, inflation will be avoided, perhaps at the expense of a temporary shortage of some commodities.

6. Appendix: Chronology in the Price History of China

1952 —

The Communist Party's General Line in the transition period is formulated.

State monopoly established on purchase of staples and marketing (grain, cotton, etc.).

Private industrial profits limited to $10\% \sim 30\%$.

Beginning in 1958 —

The Great Leap Forward Movement.

Severe shortage of market supply; sharp and persistent increase in prices.

1961 —

Eighteen categories of commodities for which prices were stabilized: food, cotton cloth, knitwear, cotton for wadding, salt, sugar, coal and charcoal, matches, stationery, rent, etc., which account for some 60% of living expenses.

High price but unlimited supply policy adopted for confectionery, restaurant dishes, famous wines, bicycles, wristwatches, knitwear, tea, granulated sugar, and imported cigarettes.

1962 —

National Price Commission of the State Council established.

Prices readjusted for metallurgical products, machinery and electrical products, coal and charcoal, timber, knitwear, bicycles, Chinese medicine, and some agricultural and sideline products.

1965 —

Price stabilization plan accomplished: retail price indices with the preceding year as 100, were 116.2 in 1961; 103 in 1962; 94.1 in 1963; 96.3 in 1964; 97.3 in 1965.

Prices in the marketplace fell drastically, high-priced goods were restored to more normal levels.

1967 —

Prices were frozen by "Notice" in 1967 during the "Cultural Revolution."

1979 —

The price for lump coal increased by 5 yuan per ton, and prices for machine parts, electrical products, and rubber goods decreased by various margins.

1981 —

Prices increased for bambooware, ironware, ceramics and porcelain, leather, tobacco, and wines, while they decreased for TV sets, wristwatches, polyester cotton fiber, and polyamide stretch socks.

1983 —

Higher rates were established for railway and waterway freight and passenger transport.

Prices declined by wide margins for synthetic fiber goods and somewhat increased for cotton fabrics.

From 1978 through 1983—

Purchase prices of agricultural products advanced 8.1% each year. The amount of increase during the five years is about three times as much as that over the 28 years prior to 1978.

1984 —

Reform of the price system was set in motion. Reform of the price system continues through 1986 and 1987.

Acknowledgment

The authors with to thank Shun-shu Wu for her patience in reading, editing, and typing the many drafts of the manuscript.

Notes

(1) It is also an expression implying the indirect agricultural contribution to the state revenue. See Chang and Lin (1985).

(2) Pork, eggs, and some vegetables are now being withdrawn from the list of subsidized goods.

(3) The original text reads "not only is no more than the necessary

labor-time used up for each specific commodity, but only the necessary proportional quantity of the total social labor-time is used up in the various groups"(see Marx, 1959, p. 620).

(4) Farm products of the first category refer to key commodities, such as grain, edible oil, and cotton, which have been under state monopoly for purchase and marketing and have a great bearing on the national economy and the people's livelihood.

Farm products of the second category refer to commodities which are fairly important to the national economy and the people's livelihood, or which are not produced in some regions, or which are in short supply, or which are produced for export. This category includes pigs, eggs, and jute.

Farm products of the third category refer to farm produce outside the first two categories, the production of which is arranged basically by the producers themselves.

(5) Industrial products produced according to mandatory planning include coal, crude oil products, rolled steel, nonferrous metals, timber, cement, electricity. basic raw materials for the chemical industry, chemical fertilizers, important machinery and electrical equipment, synthetic fibers, cigarettes and cigars, newsprint, and munitions.

In agriculture, the state will continue to set purchase or allocation quotas for such items as cereals, cotton, edible oil, cured tobacco, jute, and pigs according to their quantity, variety, and quality.

(6) The contract system was first implemented in 1979 in Chuxian Prefecture in Anhui as a way to eliminate poverty. But it has proven widely adaptable to a variety of conditions and circumstances. In Taicang County, Jiangsu province, 99.5% of the production teams have adopted the contract system. The local economy has since developed rapidly.

A peasant in Taicang is quoted as saying: "The household contract system satisfies me most in that it grants us decision-making powers. In the past, if we wanted to ask for leave or borrow some money from the production team, we had to have the permission of the team leader. But now we can arrange and do everything according to our own specialities. ...My family's net income was 1,020 yuan in 1982 when the group contract system was introduced. In 1983, when the household contract system was introduced, my family earned 2,065 yuan. This year, we've got another three new products: cultured pearls, mushrooms, and garlic. In this way we're sure we can get another 1,000 yuan."

第五部分 国际交流

参考文献

References

[1] Allen R G D. The economics of index numbers[J]. Economica,1949, 16(63).

[2] Allen R G D. Mathematical analysis for economists[M]. London: MacMillan,1938.

[3] Allen R G D. On the marginal utility of money and its application[J]. Economica,1933,13(5).

[4] Allen R G D. Some observation on the theory and practice of price index numbers[J]. Review of Economic Studies,1935,3(10).

[5] Allen R G D, Bowley A L. Family expenditure [M]. London: King,1935.

[6] Allen R G D,Hicks J R. A reconsideration of the theory of value[J]. Economica,1934(1).

[7] Arrow K J. Social choice and individual values, cowles commission [M]. New York:Wiley,1951.

[8] Baumol W J. Community indifference [J]. Review of Economic Studies,1946-1947,14(1).

[9] Baumol W J. The community indifference map: a construction[J]. Review of Economic Studies,1949-1950,17(44).

[10] Bergson A. Real income, expenditure proportionality and Frisch's "New Methods" of measuring marginal utility [J]. Review of Economic Studies,1936,4(1).

[11] Bergson A. Socialist economics [M]//Ellis H S. A survey of contemporary economics. Philadelphia:Blakiston,1949.

[12] Black J D,Mudgett B D. Research in agricultural index numbers[C].

Social Science Research Council, Bulletin No. 10, 1938.

[13] Bortkiewicz L von. Review of Gottfried Haberler's der sinn der indexzahlen[J]. Magazin der Wirtschaft, 1928, 4(11).

[14] Bortkiewicz L von. Nordic Statistical Journal, 1923, 1924, 1932.

[15] Boulding K E. Economic analysis[M]. New York: Harper, 1948.

[16] Bowley A L. Earnings and prices, 1904, 1914, 1937-1938[J]. Review of Economic Studies, 1940-1941, 8.

[17] Bowley A L. The measurement of changes in the cost of living[J]. Journal of the Royal Statistical Society, 1919, 81(5).

[18] Bowley A L. Notes on index numbers[J]. Economic Journal, 1928, 38.

[19] Bowley A L. The mathematical groundwork of economics[M]. Oxford: Clarendon, 1924.

[20] Carter C F, Reddaway W G, Stone R. The measurement of production movement[M]. Cambridge: Cambridge University press, 1948.

[21] Clark Colin. The conditions of economic progress[M]. 2nd ed. London: MacMillan, 1951.

[22] Court L M, lewis H G. Production cost indices[J]. Review of Economic Studies, 1942-1943, 10(1).

[23] Davis H S. The industrial study of economic progress[M]. Philadelphia: University of Pennsylvania, 1947.

[24] Divisia F. Economique rationelle[M]. Paris: Doin, 1928.

[25] Dresch F W. Index numbers and the general economic equilibrium[J]. Bulletin of the American Mathematical Society, 1938, 44.

[26] Edgeworth F Y. Mathematical psychics[M]. London: Paul, 1881.

[27] Edgeworth F Y. Papers relating to political economy[M]. London: MacMillan, 1925.

[28] Fisher I. The making of index numbers[M]. Cabridge: Riverside, 1922.

[29] Fisher I. The nature of capital and income[M]. New York: MacMillan, 1906.

[30] Fisher I. A statistical method for measuring "Marginal Utility" and testing the justice of a progressive income tax[M]//Economic essays contributed in Honor of John Bates Clark. New York: MacMillan, 1927.

[31] Frisch R. Annual survey of general economic theory: the problems of

index numbers[J]. Econometrica,1936,4(1).

[32] Frisch R. New methods of measuring marginal utility[M]. Tubingen: Mohr,1923.

[33] Graaff J de V. Rothbarth's "Virtual Price System" and the Slutsky Eauations[J]. Review of Economic Studies,1947-1948,15(38).

[34] Haberler G. Der sinn der indexzahlen[M]. Tubingen:Mohr,1927.

[35] Haberler G,Hagen E E. Taxes,government expenditures and national income[J]. Conference on Research in Income and Wealth,1946,8.

[36] Harrod R F. Scope and method of economics[J]. Economica,1938,48(9).

[37] Henderson A. Consumer's surplus and the compensating variation [J]. Review of Economic Studies,1940,8.

[38] Hicks J R. Consumer's surplus and index numbers[J]. Review of Economic Studies,1942,9.

[39] Hicks J R. The foundations of welfare economics[J]. Economic Journal,1939,49.

[40] Hicks J R. The rehabilitation of consumers' surplus[J]. Review of Economic Studies,1940,8.

[41] Hicks J R. The valuation of the social income[J]. Economica,1940.7.

[42] Hicks J R. The valuation of the social income—a comment of Prof. Kuznets' reflections[J]. Economica,1948,15(8).

[43] Hicks J R. Value and capital[M]. 2nd ed. Oxford:Clarendon,1946.

[44] Hotelling H. The general welfare in relation to problems of taxation and of railway and utility rates[J]. Econometrica,1938,6(7).

[45] Kaldor N. Rationing and the cost of living index[J]. Review of Economic Studies,1940,8.

[46] Kaldor N. Welfare propositions of economics and interpersonal comparisons of utility[J]. Economic Journal,1939,49(September).

[47] Keynes J M. A treatise on money, Vol. 1 [M]. New York: Harcourt,1930.

[48] Klein L R. Macroeconomics and the theory of rational behavior[J]. Econometrica,1946,14(April).

[49] Klein L R,Rubin H. A constant-utility index of the cost of living[J]. Review of Economic Studies,1947,15(38).

[50] Konus A A. The problem of the true index of the cost of living[J]. Econometrica,1939,7(1).

[51] Kuznets S. National income:a summary of finding[M]. New York: National Bureau of Economic Research,1941.

[52] Kuznets S. National income and its composition,1919-1938[M]. New York:National Bureau of Economic Research,1941.

[53] Kuznets S. On the valuation of the social income—reflections of Prof. Hicks's Article,part 1 and 2,Economica,1948,15(57-58).

[54] Lange O R. The foundations of welfare economics[J]. Econometrica, 1942,10(3-4).

[55] Leontief W. Composite commodities and the problem of index numbers[J]. Econometrica,1936,4.

[56] Lerner A F. A note on the theory of price index number[J]. Review of Economic Studies,1935,3(10).

[57] Little I M D. A critique of welfare economics[M]. Oxford:Clarendon, 1950.

[58] Little I M D. The foundations of welfare economics[J]. Oxford Economic Papers,1949,1(2).

[59] Little I M D. A note on the interpretation of index numbers[J]. Economica,1949,16(64).

[60] Little I M D. The valuation of the social income[J]. Economica,1949, 16(61).

[61] Mills F C. The behavior of prices[M]. New York:National Bureau of Economic Research,1927.

[62] Mills F C. Price data and problems of price research[J]. Econometrica,1936,4.

[63] Mitchell W C. The making and using of index numbers[M]. Washington:U. S. Bureau of Labor Statistics,Bulletin No. 284,1921.

[64] Morgan J M. Can we measure the marginal utility of money? [J]. Econometrica,1945,13(2).

[65] Mudgett B D. The cost of living index and Konus condition[J]. Econometrica,1945,8(2).

[66] Nicholson J L. Rationing and index numbers[J]. Review of Economic Studies,1942-1943,10(1).

[67] Pigou A G. Economics of welfare[M]. London:MacMillan,1920.

[68] Pigou A G. Economics of welfare [M]. 4th ed. London: MacMillan,1932.

[69] Pigou A G. Income, an introduction to economics[M]. London: MacMillan,1948.

[70] Pigou A G. Some aspects of welfare economics[J]. American Economic Review,1951,41(3).

[71] Pigou A G. The veil of money[M]. London:MacMillan,1949.

[72] Robbins L. An essay on the nature and significance of economic science[M]. London:MacMillan,1935.

[73] Robbins L. Interpersonal comparisons of utility: a comment[J]. Economic Journal,1938,48(12).

[74] Rothbarth E. The measurement of changes in real income under conditions of rationing[J]. Review of Economic Studies,1940-1941,8.

[75] Roy Rene. Les index economiques [M]//Etudes econometriques. Paris:Recueil Sirey,1935.

[76] Samuelson P A. Constancy of the marginal utility of income[M]// Lange McIntyre, Yntema. Studies in mathematical economics and econometrics in memory of Henry Schultz. Chicago: university of Chicago,1942.

[77] Samuelson P A. Evaluation of real national income[J]. Oxford Economic Papers,1950,2(1).

[78] Samuelson P A. Foundations of economic analysis[M]. Cambridge: Harvard University,1947.

[79] Samuelson P A. The gain from international trade[J]. Canadian Journal of Economics and Political Science,1939,5(5).

[80] Schultz Henry. Theory and measurement of demand[M]. Chicago: University of Chicago,1938.

[81] Scitovszky T. A note on welfare propositions in economics[J]. Review of Economic Studies,1941,9(11).

[82] Scitovszky T. A reconsideration of the theory of tariffs[J]. Review of Economic Studies,1941,9(11).

[83] Scitovszky T. The state of welfare economics[J]. American Economic Review,1951,41(3).

[84] Scitovszky T. Welfare and competition[M]. Chicago:Irwin,1951.

[85] Shoup C S. Principles of national income analysis[M]. Cambridge: Riverside,1947.

[86] Simpson P B. Transformation functions in the theory of production indexes[J]. Journal of the American Statistical Association,1951,46(254).

[87] Staehle H. A development of the economic theory of price index numbers[J]. Review of Economic Studies,1935,4(3).

[88] Staehle H. A general method for the comparison of the price of living [J]. Review of Economic Studies,1935,4(3).

[89] Staehle H. International comparison of food costs[M]//Studies and Reports No. 20,Series N. Geneva:International Labor Office,1934.

[90] Staehle H. The reaction of consumers to changes in prices and income: a quantitative study in immigrants' behavior [J]. Econometrica,1934,2.

[91] Staehle H. Report of the Fifth European Meeting of the Econometric Society[J]. Econometrica,1937,5(1).

[92] Staehle H, Joseph M F W, Lerner A P. Further notes on index numbers[J]. Review of Economic Studies,1935,4.

[93] Ulmer M J. The economic theory of cost of living index numbers[M]. New York:Columbia Unversity,1949.

[94] Viner J. Studies in the theory of international trade[M]. New York: Harper,1937.

[95] Wald A. A new formula for the index of cost of living [J]. Econometrica,1939,7(4).

[96] Wallis W A, Friedman M. The empirical derivation of indifference functions[M]//Lange McIntyre, Yntema. Studies in mathematical economics and econometrics in memory of Henry Schultz. Chicago: University of Chicago,1942.

[97] Wicksell Knut. Lectures on political economy [M]. London: Routledge,1946.

[98] Draper N R,Stoneman D M. Estimating missing values in unreplicated two-level factorial and fractional factorial designs[J]. Biometrics, 1964,20(3).(该文除在 Daniel 的半正态图纸上对估计值作检验外,可

作为本文的特殊情形。）

[99] Kempthorne O. The Design and Analysis of Experiments[M]. New York: John Wiley & Sons, Inc, 1952: 287.

[100] Kishen K. On the construction of Latin and Hyper-Graeco-Latin cubes and hypercubes[J]. J Indian Soc Agr Stat, 1949: 22-23.

[101] Sclove S L. On missing value estimation in experimental design models[J]. American Statistician, 1972, 26（2）25-26.（据文摘《Abstract Service: Quality Control and Applied Statistics》1972 年第 6 期报道，该文可能有原理上类似于本文中关于等效性的证明，但作者未获读原著。）

[102] Mann H B. 试验的分析与设计[M]. 张里千, 等译. 北京: 科学出版社, 1964.

[103] 李士琦. 电渣冶炼工艺参数对过程脱硫率的影响——实验设计法应用一例[J]. 数学的实践与认识, 1972(3): 35-39.

[104] Feller William. An introduction to probability theory and its applications Vol. 1[M]. 3rd. New York: John Wily & Sons, Inc, 1968.

[105] Fisher R A. Statistical methods and scientific inference[M]. London: Oliver and Boyd, 1956.

[106] Kiefer J. Statistical inference [M]//The mathematical sciences. Massachusetts: Mit Press, 1969.

[107] Basu D. On sampling with and without replacement. Sankhya, 1958, 20: 287-294.

[108] Bowley A L. Measurement of the precision attained in sampling. Bull. Int Stat Inst, 1926, 22: 1-62.

[109] Chang Weiching. Statistical theories and sampling practice. 1969.

[110] Cochran W. Sampling Techniques, Wiley, 1977.

[111] Ericson W A. (1969). Subjective Bayesian Models in sampling finite populations: stratification, In New Developments In Survey Sampling. N. L. Johnson and H. Smith(eds.), pp. 326-357.

[112] Godambe V P. Some aspects of the theoretical developments in survey sampling[J]. Ibid. 1969: 27-58.

[113] Goodman Leo. Snowball sampling. Ann Math Statist[J]. 1961, 32: 148-170.

[114] Hansen M H, Hurwitz W N. On the samplig from finite populations

[J]. Ann Math Statist,1943,14:333-362.

[115] Harley H O,Rao J N K. A new estimation theory for sample survey. I. Biometrika [J], 1969, 55: 547-557. II. In New Developments In Survey Sampling[J]. 1969:147-169.

[116] Kalbfleisch J D, Sprott D A. Applications of likelihood and fiducial probability to sampling finite populations [J]. Biometrika, 1969: 358-389.

[117] Kempthorne O. Some Renarks on statistical inference in finite sampling[J]. Biometrika,1969:671-695.

[118] Laplace P S. Theorie Analytique Des Probabilites,1812.

[119] Hansen M H,Madow W G. Some important events in the historical development of sample surveys,1969.

[120] Neyman J. (1934). On the two different aspects of the repres entative method[J]. J Roy Statist Soc. Ser 1934,97:558-625.

[121] Royall R M. On finite populaton sampling theory under certain linear regression models[J]. Biometrika,1970,57:377-387.

[122] Seng Y P. Historical Survey of the developm of sampling theories and practice[J]. J Roy Statist Soc A,1951,114:214-231.

[123] Stephan F F. Three extensions of sample survey technique: hybrid, nexus,and graduated sampling[M]. In New Developments In Survey Sampling,1969:81-104.

[124] Zarkovic S S. Note on the history of Sampling methods in Russia[J]. J Roy Statist Soc A,1959,119:336-338.

[125] Zarkovic S S. A supplement to "Note on the history of sampling in Russia[J]. J Roy Statist, Soc A,1962,125:580-582.

[126] 《现代工程数学手册》编写组. 现代工程数年学手册第Ⅳ卷[M]//林少宫. 第61卷:正交设计法. 武汉:华中理工大学出版社,1987.

[127] Wooldridge J M. Introductory econometrics—A modern approach [M]. Mason:South-Western College Publishing,2000.

[128] Bienens H J, Swanson N R. The economic consequences of the "ceteris paribus" condition in economic theory [J]. Journal of Econometrics,2000,95(2).

[129] Fisher R A. The arrangement of field experiments[J]. Journal of Ministry of Agriculture,1926,33.

[130] Ashenfelter O, Kreuger A B. Estimates of economic return to schooling from a new sample of twins[J]. American Economic Review,1994,84:1157-1173.

[131] Blackburn M,D. Neumark Unobserved ability,Efficiency wages,and interindustry wage differentials[J]. Quarterly Journal of Economics,1992,107:1421-1449.

[132] Heckman J. The common structure of statistical models of truncation, sample selection, and limited dependent variables and a simple estimator for such models[J]. Annals of Economie and Social Measurement,1976,5:475-492.

[133] McFadden D. 怎样量化环境损坏或改善的经济价值[A]. 田国强. 现代经济学与金融学前沿发展[C]. 北京:商务印务馆,2002:52-62.

[134] Stiglitz J E,Walsh c E. Economics[M]. New York:W. W. Norton & Company Inc,2000.

[135] Wooldridge J M. Introductory econometrics—A modern approach [M]. Mason:South-Western College Publishing,2000.

[136] 林少宫. 正交实验设计在微观计量经济中的应用——政策(或事件)评价法[J]. 湖北省现场统计研究会通讯,2002(4).

[137] 《现代工程数学手册》编写组. 现代工程数年学手册第Ⅳ卷[M]//林少宫. 第61卷:正交设计法. 武汉:华中理工大学出版社,1987.

[138] Branscomb Lewis M. Infromation technology as an economic equalizer[J]. Economic Impact,1986,56:44.

[139] Chang Peikang, Lin Shaokung. China's modernization: stability, efficiency and the price mechanism [M]//Dutta M. Asia-Pacific economies: promises and challenges, Part A. Greemwich CT: JAI press,1987:103-118.

[140] Chinese Academy of Social Science-Institute of Economics. Economic Studies,1985,43.

[141] Deng Xiaoping. Jianshe you zhongguo tese de shehuizhuyi(enlarged edition). Beijing:The people's press,1987.

[142] Du Wenbai. Some aspect of utilizing foreign capital[J]. International Trade Journal,1984(2).

[143] Fusfeld H I,Haklisch C S. Industrial productivity and international technical cooperation[M]. New York:Pergamon press,1982.

[144] Government of China. Provisions encouraging foreign investment in China[M]. Beijing:Government printing office,1986.

[145] Guang Ming Ri Bao. 1987-5-28,1987-6-15.

[146] Heller Peter B. Technology transfer and human values[M]. Lanham, NY:Unversity press of America,1985.

[147] McLean Murray. Australia/China relations-political and economic perspectives[M]//China's Trade with Other Pacific Rim Nations, U. S. Congress Hearing No. 98-163. Washington, DC: U. S. Government Printing Office,1985.

[148] National statistical bureau of China. Almanac of Statistics-China [M]. Beijing:Statistics publishing office,1986.

[149] Nau Henry R. Competition vs coordination in technology transfer [J]. Economic Impact,1986,58.

[150] Sahal D. The transfer and utilization of technical knowledge[M]. Lexington,MA:Lexington books,1982.

[151] U. S. Embassy in Beijing. Economic Background No. 100[R]. 1984.

[152] The World Bank. Economic structure in international perspective (annex 5 China: Long-term development, issue and options)[Z]. 1985,38.

[153] Almanac of Statistics-China. Beijing: State statistical Bureau,1984: 426,551.

[154] Central Committee, CPC. Decision of the Central Cmmittee of the Communist Party of China on reform of the economic structure[C]. Adopted by 12th Central Committee of the CPC at its Third Plenary Session,1984,Oct. 20.

[155] Chang Peikang, Lin Shaokung. China's economic readjustment and trade perspective[A]. Dutta. Studies in United States-Asia Economic Relations[C]. Durham,NC:Acorn press,1985.

[156] Chinese energy economics research group. Discussions on energy prices[J]. Price theory and practice,1983(5):1.

[157] Editorial department. The glorious achievements of the price work for the past thirty-five years[J]. Price theory and practice, 1984 (5):61.

[158] Marx Karl. Capital Vol 3[M]. Moscow:Foreign language pubishing house,1959:620.

[159] Qiao Yongzhang. Evolution, the present state and the role of price subsidization[J]. Price theory and practice,1983(1):1.

后记
Postscript

　　两年前，我们准备筹划编辑出版一本林少宫先生的文集，以此纪念先生。这项工作主要由林先生的长子林子美承担，也得到林先生大弟子田国强教授的大力支持。国强师兄专门为文集撰写了序言。子美大哥花了大量时间和精力，搜集整理了包括林少宫先生的博士论文在内的不少珍贵文献，其中不少未在国内出版过。由于收集整理和资料考证工作十分繁琐，学院特地委派青年教师郭宁予以协助，并负责与出版社编辑之间做好有关协调事务。原本计划当年出版付印，但后来不断发现新的资料需要核实考证，再加上随后武汉突如其来遭受新冠肺炎疫情的冲击，编校出版工作只好顺延。由于出版篇幅受限，这本文集最终精选了林少宫先生有代表性的论文，并决定以电子阅读件方式附上收集整理的其他文献（阅读网址：http://eco.hust.edu.cn/info/1269/12854.htm）。现在总算是正式出版付印了，我们要对完成这项工作付出辛勤劳动的各位朋友表示最衷心的感谢！今年恰逢我校经济学科创办40周年，明年也即将迎来林少宫先生百年诞辰。因此，我们能及时出版《林少宫文集》，具有极其重要的纪念意义。

　　林少宫先生是我国著名的数理统计学家和计量经济学家。他的学术成就主要有三点：一是推动了数理统计学在中国的普及和发展；二是研究并推广了正交试验设计方法，为社会带来了巨大的经济效应；三是倡导并推动了数量经济学在我国的建立和发展。在半个多世纪的学术生涯中，他不仅在理论上造诣深厚，出版学术著作10余部，发表学术论文近百篇；而且在数理统计的实际应用方面推广了正交实验设计方法，为国家现代化建设创造了巨大的经济效益。同时，林先生还培养了数十名优秀的研究生，其中田国强、艾春荣、谭国富、宋敏等，都已经成为知名经济学家，并在国际学术界崭露头角。由于林先生在理论、方法、应用和教学各个方面取得的杰出成就，其事迹先后被收入《世界（教育界）名人录》和《中国世纪专家传略》等传记丛书。而这次收集的论文，有一些内容是首次与广大读者见面,因此,文集的出版可以让我们更全面地了解先生的为学之道。

林先生也是我校(原华中工学院,后华中理工大学,现华中科技大学)经济学科的开创者之一。我校经济学科有两个渊源:一是1981年成立的经济研究所,创始所长是张培刚教授,这是在社会科学部政治经济学教研室基础上组建的,建所后便开始招收经济学专业方向研究生;另一个渊源是1982年学校在数学系成立的数量经济学研究所,创始所长是林少宫先生,并开始在数学学科招收数量经济学专业方向研究生。1985年学校汇集经济所、数量所、管理工程系等单位组建经济管理学院,张培刚教授任名誉院长,林少宫先生任院长。1994年学校组建独立的经济学院后,林先生担任顾问积极支持学院发展,并担任数量经济与金融研究中心主任。经过多年努力,如今我校已经形成完整的经济学科体系,包含经济学、金融学、经济与贸易、金融工程、经济统计学和国际商务六个本科专业,理论经济学、应用经济学硕士、博士授予点和博士后流动站,此外,还有金融和国际商务专业硕士点。可以说,正是有了张培刚、林少宫老一辈大师的创办和带领,我们才拥有两个具有特色的学科领域:发展经济学和数量经济学,也才有了今天的发展壮大和成就。

借此机会,我也想简要回忆一下我与林先生的交往片段。早在大学本科期间我就读华中工学院计算机系,曾学习过林先生编写的《基础概率与数理统计》。1988年我开始在张培刚老师的名下攻读经济学硕士研究生,正好与数量经济学专业方向的研究生为室友,经常听他说起他导师林少宫先生,要求他们读数量经济学原著和最新英文文献并定期汇报研讨,不久我便开始旁听研讨会,一起与他们分享经济建模和经济统计问题的研讨。后来我一直在学校工作,这样就有更多机会当面向林先生单独请教。记得1993年,中国留美经济学会组织编写出版了14本"市场经济学普及丛书",其中有一本就是林少宫、李楚霖老师合著的《简明经济统计与计量经济》。当时,我正在武汉大学攻读经济学博士,正好有问题向林先生请教,林先生特意与我探讨了完备的经济学知识体系建构的重要性,并专门送了我一本他的新著,鼓励我文理交叉不断探索,让我深受教益。第二年,邹恒甫教授回到武汉大学创办高级研究中心,义务为经济学院、数学学院的研究生系统讲授高级宏观、微观经济学,特意推荐林少宫、李楚霖老师合著的《数理经济学导论》作为入门书,并委托我到华中理工大学出版社书库采购100本(每本1.5元,共150元,由邹教授出钱)免费送给参加上课的同学。事后,我将此事告诉了林先生。林先生十分感慨,非常认同邹教授的办学义举,也鼓励我们的校友回校举办现代经济学前沿讲座和经济学-数学双学位实验班。林先生说,经济学需要文理贯通人才,我们的教师也好、研究生也好,需要文就补文,需要理就补理。中国经济问题埋藏着深层"文"的问题,数量经济学有许多"理"的成分,要求我们偏理的要补文,偏文的要补理。在后来的交往中,林先生对我的教导很多、鼓励很大。有一次,我们在探讨经济学说史如何有效讲授时,我提出"应从当代诺贝尔经济学奖得主的学术贡献和现实影响,反思和追溯经济思想的形成与发展",他十分认同,并予以充分肯定和热情鼓励。又一次,当我登门请教最新计量经济学方法时,他将刚刚看到并整理的关于面板数据估计的手稿给我学习。林先生八十高龄时仍十分关注学术前沿发展,亲自主编了《微观计量经济学要义》,当时我正在学校支持文科办公室工作,力推此书入选文科丛书优先出版。正如林先生主张:我们学院要向有特色研究型学院方向发展,应注重发展经济学与数量经济学的长远结合,并以中国经济问题为融合点,作为我们可以考虑树立的一个标识。应该说,正是有了两者的有机结合,我们学院的学科发展才有了鲜明的特色,也迎来了大发展。

斯人已去,风范长存!我衷心希望通过本书能很好传承老一辈学者的治学精神,更能激励青年一代学子们"无论如何都应为祖国振兴而效力"!

张建华

华中科技大学经济学院院长、教授

2021 年 8 月 21 日

图书在版编目(CIP)数据

林少宫文集/林少宫著.—武汉:华中科技大学出版社,2021.12
ISBN 978-7-5680-7789-7

Ⅰ.①林…　Ⅱ.①林…　Ⅲ.①数学-文集　Ⅳ.①O1-53

中国版本图书馆 CIP 数据核字(2021)第 260503 号

林少宫文集
Lin Shaogong Wenji

林少宫　著

策划编辑:	周晓方　陈培斌　宋　焱
责任编辑:	肖唐华
封面设计:	原色设计
责任校对:	曾　婷
责任监印:	周治超
出版发行:	华中科技大学出版社(中国·武汉)　　电　话:(027)81321913
	武汉市东湖新技术开发区华工科技园　　邮　编:430223
录　排:	华中科技大学惠友文印中心
印　刷:	湖北金港彩印有限公司
开　本:	710mm×1000mm　1/16
印　张:	17.75　插页:12
字　数:	298 千字
版　次:	2021 年 12 月第 1 版第 1 次印刷
定　价:	168.00 元

本书若有印装质量问题,请向出版社营销中心调换
全国免费服务热线:400-6679-118　竭诚为您服务
版权所有　侵权必究